THE MEANING
OF SCIENCE

THE MEANING OF SCIENCE

An Introduction to the Philosophy of Science

Tim Lewens

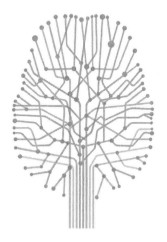

BASIC BOOKS
A Member of the Perseus Books Group
New York

Copyright © 2016 by Timothy Lewens

Published in Great Britain by Allen Lane, The Penguin Press

Published in the United States by Basic Books,

A Member of the Perseus Books Group

Books published by Basic Books are available at special discounts for bulk purchases in the United States by corporations, institutions, and other organizations. For more information, please contact the Special Markets Department at the Perseus Books Group, 2300 Chestnut Street, Suite 200, Philadelphia, PA 19103, or call (800) 810-4145, ext. 5000, or e-mail special. markets@perseusbooks.com.

Designed by Jeff Williams

Library of Congress Cataloging-in-Publication Data
Names: Lewens, Tim.
Title: The meaning of science : an introduction to the philosophy of science / Tim Lewens.
Description: New York : Basic Books, [2015] | Includes bibliographical references and index.
Identifiers: LCCN 2015039234 | ISBN 9780465097487 (hardcover) | ISBN 9780465097494 (e-book)
Subjects: LCSH: Science—Philosophy.
Classification: LCC Q175 .L477 2015 | DDC 501--dc23 LC record available at http://lccn .loc.gov/2015039234

10 9 8 7 6 5 4 3 2 1

For Rose and Sam

Contents

Acknowledgments

My first debt is to Laura Stickney at Penguin, who was kind enough to ask me to write this book. She has been an enthusiastic and patient editor, with a gratifyingly delicate touch. I also owe thanks to many colleagues and friends. Anna Alexandrova, Riana Betzler, Adrian Boutel, Andrew Buskell, Christopher Clarke, Chris Edgoose, Beth Hannon, Stephen John, and Huw Price all read the draft manuscript in its entirety. My wife Emma Gilby's comments were especially valuable, and I'm otherwise indebted to her for innumerable reasons. For advice on individual chapters and shorter sections I'm grateful to Jonathan Birch, Hasok Chang, Helen Curry, Dan Dennett, Jeremy Howick, Nick Jardine, Lisa Lloyd, Aaron Schurger, and Charissa Varma. For other forms of education, support, and encouragement I would like to thank Tamara Hug, Christina McLeish, Helen Macdonald, Hugh Mellor, Louisa Russell, and David Thompson. I'm indebted to the University of Cambridge and to Clare College for allowing me time to write this book, to my new colleagues at the Centre for Research in the Arts, Social Sciences, and Humanities (especially Simon Goldhill and Catherine Hurley) for providing an exciting environment in which to finish the book, and to the European Research Council (grant number 284123) for funding much of the research that contributed to it. I must also thank the many students I have worked with at Cambridge, and who have made me think hard about what the philosophy of science is, and why it matters. In putting together a book that aims to introduce people to this subject, I have found myself thinking again about Peter Lipton, an exemplary teacher and a friend I still miss.

This book is dedicated to my children, Rose and Sam. I couldn't honestly say that without them the book wouldn't have been possible. But it would have been different, it would probably have been worse, and I wouldn't have enjoyed writing it half as much.

A Note for Readers

The chapters in this book are all more or less self-contained, so they can be read in any order. Each one finishes with a short guide to further reading, which picks out some accessible books for those who wish to find out more about the topics discussed. Most readers can safely ignore the book's many endnotes. They indicate the sources for the facts, arguments, and claims mentioned in the main text.

Introduction

The Wonder of Science

The achievements of the sciences are extraordinary. They have produced explanations for everything from the origins of human culture to the mechanisms of insect navigation, from the formation of black holes to the workings of black markets. They have illuminated our moral judgments and our aesthetic sensibilities. Their gaze has fallen on the universe's most fundamental constituents and its very first moments. They have witnessed our intimate private activities and our collective public behaviors. Their methods are so compelling that they can command consensus even when dealing with events that are invisible or intangible, in the distant past or the distant future. Because of this, the sciences have alerted us to some of the most pressing problems facing humanity, and the sciences will need to play central roles if these problems are to be solved.

This book—an introduction to the philosophy of science—steps back from the particular achievements of the sciences to ask a series of questions about the broad significance of scientific work. It is a book for anyone with an interest in what we mean by "science," and in what science means for us. It does not assume any scientific knowledge, nor does it assume any familiarity with philosophy.

The philosophy of science, like all branches of philosophy, has existed since the time of the ancient Greeks. And like all branches of philosophy, it has a mixed reputation. The charismatic American physicist Richard Feynman—a recipient of the Nobel Prize for physics in 1965—had little patience for the subject, allegedly remarking that "philosophy of science is about as useful to scientists as ornithology is to birds."[1]

Feynman's words—assuming he really said them—were ill chosen. Ornithology is useless to birds because birds cannot understand it. If a bird could only learn what ornithologists know about how to recognize a cuckoo chick in its brood, then that bird could save itself a lot of misguided effort. Of course, Feynman didn't mean to suggest that philosophy was too complicated for scientists to comprehend; he just didn't see any evidence that philosophy could contribute to scientific work.

There are many good ways to respond to this challenge. One comes from a physicist whose stature is even greater than Feynman's. In 1944, Robert Thornton, freshly qualified with a PhD in the philosophy of science, began teaching modern physics to students at the University of Puerto Rico. He wrote to Albert Einstein for advice. Should he introduce philosophy into his physics course? Einstein wrote back with an unequivocal "yes." "So many people today," he complained, "and even professional scientists—seem to me like somebody who has seen thousands of trees but has never seen a forest." Einstein went on to describe the antidote to this myopia:

A knowledge of the historic and philosophical background gives that kind of independence from prejudices of his generation from which most scientists are suffering. This independence created by philosophical insight is—in my opinion—the mark of distinction between a mere artisan or specialist and a real seeker after truth.[2]

For Einstein, the value of the philosophy of science, in combination with the history of science, lay in its ability to liberate the investigator's imagination.[3]

We will see in this book that the sciences have been admirably ambitious in bringing their methods to some of the most profound topics the world presents us with. Psychologists, evolutionists, and neuroscientists have grappled, for example, with the nature of ethics and the reality of free choice. Once they venture down these investigative pathways, it is impossible for them to avoid engagement with philosophy. Scientists cannot make plausible pronouncements about the repercussions of evolutionary theorizing for human morality, they cannot assess the fate of free will in the face of work in neuroscience, unless they have well-formulated views about what morality, or freedom of the will, involve. In other words, whether they like it or not, scientists end up running into exactly the same conceptual issues that have puzzled philosophers for centuries.

This does not mean that philosophers have nothing to learn when scientists begin to colonize territory that has traditionally belonged to the humanities. On the contrary, recent philosophical work on topics like morality and free will has been greatly enriched by its interactions with the best scientific research on evolution, the mind, and social behavior. In areas like these, philosophy and the sciences have repeatedly come together in constructive ways. They have learned from each other.

We should not suppose that the value of the philosophy of science is fully measured by the degree to which it helps scientists. It also has general cultural significance. The sciences look everywhere, but do they see everything? Will they eventually teach us all that is worth knowing? Or are there alternative forms of understanding that must be arrived at in other ways, perhaps by engaging with works of literature, perhaps by abstract reflection? Philosophical questions like these concern the reach of science, and they help us to understand how the sciences and the arts make different kinds of contributions to human knowledge.

The philosophy of science also has direct political relevance. We cannot ascertain how governments should respond to threats from climate change without first determining how we should reason when our evidence is uncertain and when the stakes are momentous. We cannot decide whether homeopathic treatments should be funded by public health budgets without asking about the markers of genuine science and the markers of pseudoscientific quackery. We cannot assess how democratic states should make use of technical scientific advice without inquiring about whether apparently neutral pieces of scientific information already come laden with moral and political values.

It turns out, in other words, that the issues addressed by the philosophy of science—the issues we will explore in this book—matter in the most practical ways, for the most important questions of all.

Part One

What We Mean by *Science*

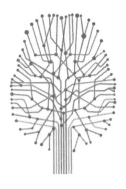

Chapter One

How Science Works

Science and Pseudoscience

There are sciences. Physics is one, chemistry another. There are also disciplines that involve the generation of knowledge and insight, but that few of us would immediately think of as sciences. History and literary studies are examples. All this is fairly uncontroversial. But there are cases where we are unsure about what counts as science, and these cases are sometimes politically and culturally explosive.

Consider the trio of economics, intelligent-design theory, and homeopathy. The only thing that unites these three endeavors is that their scientific status is regularly questioned in ways that provoke stormy debate. Is economics a science? On the one hand, like many sciences, it oozes both mathematics and authority. On the other hand it is poor at making predictions, and many of its practitioners are surprisingly blasé when it comes to finding out about how real people think and behave.[1] They would rather build models that tell us what would happen, under simplified circumstances, if people were perfectly rational.

So perhaps economics is less like science, and more akin to *The Lord of the Rings* with equations: it is a mathematically sophisticated exploration of an invented world not much like our own.

The theory of intelligent design has been promoted by organizations like the prominent US think tank The Discovery Institute, and developed by theorists including the biochemist Michael Behe and the mathematician/philosopher William Dembski. It aims to compete with the theory of evolution as an account of how species became well adapted to their surroundings. It suggests that some organic traits are too complex to have been produced by natural selection, and that they must instead have been produced by some form of intelligent oversight: perhaps God, perhaps some other intelligent agent. The theory is positioned as a science by its adherents, but many sensible commentators think that this is merely an attempt to insert a contentious interpretation of religion into schools, and that—understood as a piece of science—the theory is hopeless.[2]

Mainstream doctors sometimes value homeopathic remedies, in spite of the fact that their track record of validation by large-scale clinical studies is poor. One camp says that these are quack treatments with no scientific credentials, whose apparent effectiveness derives from nothing more than the placebo effect.[3] Another camp tells us that the dominant method by which scientific investigation establishes the credentials of medical interventions gives us generic wisdom regarding what works in typical circumstances for average patients, but that this approach ignores the need for doctors to prescribe what is right for a unique individual in idiosyncratic circumstances.[4]

These questions about the markers of proper science are important. They affect the power held by people whose advice can determine our financial and social well-being; they affect what our children are taught at school; they affect what forms of research our tax contributions can be used to fund and how our doctors advise that we maintain our health. These questions are also old: while today we might be concerned by the scientific status of enterprises like economics, intelligent design, and homeopathy, previous thinkers have been troubled by the scientific status of Marxism, psychoanalysis, and even evolutionary biology. What we need, it seems, is a clear account of what makes something a science and what makes something pseudoscience. What we need, it seems, is Karl Popper.

Sir Karl Popper (1902–1994)

It is still the case that if you ask a scientist to reflect on the general nature of science, you will probably be referred to the pronouncements of Karl Popper. Popper was born in Vienna in 1902, a time when Viennese cultural life was blessed with an extraordinary richness. He began attending the University of Vienna in 1918, where he exposed himself to the conspicuous intellectual movements of the moment. He became involved with left-wing politics, he adopted Marxism for a time, he listened to a lecture on relativity theory by Einstein, and he briefly served as a volunteer social worker in one of the clinics founded by psychotherapist Alfred Adler. In 1928 Popper was awarded a PhD in philosophy, and by 1934 he had published his first book, *Logik der Forschung* (later translated into English as *The Logic of Scientific Discovery*).[5] The broad conception of scientific progress laid out in that book would remain more or less intact in Popper's thinking until his death.

Popper—whose parents were of Jewish origin—was forced to leave Vienna in the 1930s. He moved to New Zealand, to a position at the University of Canterbury in Christchurch, where he spent nearly ten years before moving back to Europe. In 1946 he was offered a post at the London School of Economics (LSE), which he held until his retirement. The philosopher of science Donald Gillies, who first met Popper at the LSE in 1966, recently painted a lively picture of some of Popper's idiosyncrasies:

> Waiting in the lecture hall for Popper to appear was not without some amusement, because a ritual was always performed before the great man entered the door. Two of Popper's research assistants would come into the room before him, open all the windows, and urge the audience on no account to smoke, while writing: NO SMOKING on the blackboard. Popper had indeed a very strong aversion to smoking. He claimed that he had a very severe allergy to tobacco smoke, so that inhaling even a very small quantity would make him seriously ill. When his research assistants had reported back that the zone was smoke-free, Popper would enter the room.[6]

Gillies goes on to explain that when Popper went to a specialist in allergies, the expert was unable to find any evidence of an allergy to tobacco smoke: "Popper's comment on the result was: 'This goes to show how backward medical science still is.'"[7]

Perhaps the high point of Popper's reputation came in the late 1960s and early 1970s. He was knighted in 1965, and around this time a string of distinguished scientists described his work in tones of dazzled admiration. Sir Peter Medawar, a Nobel Prize winner for medicine, said simply: "I think Popper is incomparably the greatest philosopher of science that has ever been." Sir Hermann Bondi, mathematician and cosmologist, took the view that "there is no more to science than its method, and there is no more to its method than Popper has said."[8]

Some more of Donald Gillies's recollections make it clear that Popper could provoke exasperation, as well as admiration. On

Tuesday afternoons, the London School of Economics hosted the "Popper Seminar," where visiting speakers were invited to present their philosophical views. In a standard academic seminar of this kind, the speaker might talk unmolested for thirty or forty minutes, before the chair invites questions from the audience. At the Popper Seminar, things were different:

> Usually the speaker was allowed to talk for only about 5 to 10 minutes before he was interrupted by Popper. Popper would leap to his feet, saying that he wanted to make a comment, and then talk for 10 to 15 minutes. A typical intervention by Popper would have the following form. He would begin by summarising what the speaker had said so far. Then he would produce an argument against what the speaker had said, and he would usually conclude with a remark like: "Would you agree then that this is a fatal objection to your position?" As can be imagined such an attack would often have a very disconcerting effect on the visiting speaker.

Gillies adds: "It is easy to see that while, from Popper's point of view, his seminar could be seen as a perfect example of 'free criticism,' it could have seemed to the speaker very much like a session of the committee on un-Popperian activities."[9]

"What Is Wrong with Marxism, Psychoanalysis, and Individual Psychology?"

Popper's basic outlook on science derived from two underlying sources of discomfort. He had grown up in a place and a time of intoxicating intellectual excitement. He recalled that "after the collapse of the Austrian Empire there had been a revolution in Austria: the air was full of revolutionary slogans and ideas, and new and often wild theories."[10] Various grand intellectual systems of exceptional ambition—Einstein's relativity theory, Karl Marx's theory of history, diverse psychoanalytic understandings of the mind—were in common currency. And yet, Popper felt that there was a deep difference between relativity theory, which he venerated, and (for example) psychoanalytic theory, of which he was deeply suspicious.

He set himself the task of clarifying his intuition: "What is wrong," he asked himself, "with Marxism, psychoanalysis and individual psychology? Why are they so different from physical theories, from Newton's theory, and especially from the theory of relativity?"[11] Popper's view was that while Einstein had proposed a theory that was heroically vulnerable to destruction if experiment should show it false—and yet it had nonetheless enjoyed spectacular experimental successes—the psychoanalytic theory of mind was couched in such noncommittal terms that it was immune to experimental refutation. "I felt," he said, "that these other theories, though posing as sciences, had in fact more in common with primitive myths than with science; that they resembled astrology rather than astronomy."[12]

The problem with the predictions of newspaper astrology columns is not that they don't come true: the problem is that they are formulated in such a way that they cannot but come true, and because of that they say nothing of value. My own *Daily Mail* horoscope for the week I write these words tells me: "You have faced more downs than ups in recent weeks, but now things are about to change. With both the Sun and Venus, planet of harmony, entering your birth sign this week, you can stop worrying about the past and start planning for the future. This is also the time to bring to the boil something that has been on the back burner for too long."[13] How often would we think it sensible to advise someone to "stop planning for the future and start worrying about the past"? If something has indeed been on the back burner for "too long," doesn't that make it trivially true that now is the time to address it? And how on earth are we supposed to quantify the relative number of "ups" and "downs" we have had over the course of weeks? It is hard to see how we can argue with any of these platitudes.

Similarly, Sigmund Freud recalled how a female patient, whom he described as "the cleverest of all my dreamers," told him of a dream that seemed to refute his own theory of wish fulfillment. That theory says that in dreams our wishes come true:

> One day I had been explaining to her that dreams are fulfil-ment of wishes. Next day she brought me a dream in which she was traveling down with her mother-in-law to the place in the country where they were to spend their holidays together. Now I knew that she had violently rebelled against the idea of spending the summer near her mother-in-law and that a few days earlier she had successfully avoided the propinquity she dreaded by engaging rooms in a far distant resort. And now her dream had undone the solution she wished for: was not this the sharpest possible contradiction of my theory that in dreams wishes are fulfilled?[14]

This woman dreamed, not of something she wanted to do, but of something she abhorred: a holiday with her mother-in-law. In spite of apparent refutation, Freud argued that his theory was intact: "The dream showed I was wrong. *Thus it was her wish that I might be wrong, and her dream showed that wish ful-filled.*"[15] A dream that seems to jar against Freud's theory is ex-plained away with the argument that the woman wanted Freud to be wrong, and the dream allowed this desire to be fulfilled. It is hard not to share Popper's discomfort in the face of examples such as these. Freud's ability to cook up interpretations of the evidence that bring it into line with his theory hardly seems a strength of his psychoanalytic approach; instead, the elastic ability of his theory to stretch around whatever evidence may confront it seems more like a weakness.

The Problem of Induction

One set of Popper's concerns derived from this urgent sense that we should be able to give a "criterion of demarcation'" that will tell us how to sort science from pseudoscience. The second set of concerns came instead from Popper's deep skepticism of what philosophers call *inductive inference.* The eighteenth-century Scottish philosopher David Hume is usually credited with being the first to pose what we now call "the problem of induction." To understand this problem, we first need to understand the nature of *deductive*—as opposed to inductive—inference.

Suppose you know that all badgers are mammals, and you know that Brock is a badger. Given these premises, you can safely conclude that Brock is a mammal. This inference is *deductively valid,* meaning that it is strictly impossible for the premises of the inference to be true, and the conclusion false. There is no way that we could imagine circumstances under which all badgers are mammals, Brock is a badger, and yet Brock is not a mammal. Good deductive inferences deal in certainty: their premises ensure their conclusions. Because of this, deductive inferences are often trivial or unproductive: there is a sense in which, armed with the knowledge that Brock is a badger, and that all badgers are mammals, you are simply spelling out a self-evident consequence of those pieces of information when you go on to conclude that Brock is a mammal.

Inductive inferences are different. Suppose you have invented a new drug—let's call it *Veritor*—and you want to find out if it is safe. You test it on ten thousand people, and over a period of many months you do not detect adverse side effects in any of them. The people you choose to test are not all the same: you make sure you have tried the drug out on men, women, people of different ages, and people from different countries. Now suppose you ask the question: "Given that everyone tested so far has experienced no side effects, should we expect Colin, who has never taken the drug before, to experience adverse side effects?" I doubt that anyone would say we can be absolutely sure that Colin will be fine, but most people would say that it is reasonable to expect, on the basis of our extensive testing of the drug, that Colin will probably experience no adverse reaction.

Inferences of this sort are potentially far more valuable than deductive inferences, for they promise to generate important new knowledge. By looking at large, but limited, samples of people, we presume that we can make fairly reliable predictions about how other people are likely to react. Our practices of drug testing—and almost all other forms of knowledge-generation—seem to presuppose that it is reasonable to generalize in this way, via extrapolation from a limited number of observed instances. What makes this presupposition reasonable? The challenge inherited from Hume is to provide a justification for inductive inferences of this sort.

An inductive inference can be defined as any pattern of argument that we regard as reasonable, but which does not claim deductive validity. Our inference about Colin is not deductively valid, and it does not pretend to be. It does not deal in certainty, for clearly it is possible for ten thousand people to have experienced no side effects and for poor Colin to be the first to react badly. Such circumstances can easily be imagined without contradiction—perhaps Colin has an exceptionally rare genetic mutation—and it is partly because of this that we cannot be sure that Colin will be free from adverse reactions. Even so, we do take the view that our evidence, derived from testing thousands of people, makes it reasonable to conclude that Colin is unlikely to suffer adverse reactions. What makes this inductive inference reasonable?

We might try to justify our inference by appealing to further pieces of scientific research. For example, we might point out that for Colin to react in a way that is different from every one of the ten thousand individuals we tested previously, Colin would need a very unusual sort of body. We might go on to claim that it is reasonable, although not a certainty, to think that Colin's body is typical, because human conception and development run along well-understood lines. The processes by which human bodies are typically made have been studied in painstaking detail by physiologists and developmental biologists, and this research gives us knowledge about how Colin's body probably works, what constitutes his genetic makeup, and so forth.

This appeal to background scientific knowledge does not solve Hume's problem. It simply reveals the depth of our reliance on inductive inference. Scientists have studied a limited number of embryonic unfoldings—in humans, other mammals, and various additional species. We assume that the processes that went into the construction of Colin were most likely similar to the processes that have been observed in the laboratory. Our inference about Colin's constitution is based on extrapolation, and Hume's challenge was to explain why this form of extrapolation should be thought reasonable.

The problem of induction can be put forward as a pithy dilemma: we want to know what, if anything, makes it sensible to extrapolate from a limited sample to a broader generalization. We cannot try to answer this by claiming deductive validity for our inference, for there is evidently no contradiction in the claim that our new case is freakish, and utterly unlike what we have encountered before. But if instead we try to answer our question by pointing to scientific knowledge, or even to the past successes of previous inductive inferences, it seems we are just offering yet more instances of the very extrapolations we are trying to justify. Either way, our initial challenge—what makes extrapolation reasonable—remains unanswered.[16]

It is time to bring our discussion of induction back to Popper. Faced with a tricky crossword puzzle, we know there must be a solution even if we aren't quite sure what that solution is. Most philosophers—but not Popper—think of the problem of induction as a puzzle in this same sense: they have had a devilishly difficult time figuring out what the answer to Hume's challenge is, but they are confident there must be a good answer. After all, no one gets by in day-to-day life without induction. We are all convinced that it is better to attempt to leave a room by opening a door than by walking through the wall. We are so convinced because we extrapolate from past experience of bumps, bruises, and the frustration caused by walking into solid surfaces. When our financial advisors remind us that past successes of investments may not indicate their likely future performance, we accept their warnings because we know how often healthy funds have crashed in the past. Even here, we project past patterns into the future, and we think these extrapolations are sensible.

Popper is an outlier in the debate over induction. He understood Hume to have shown that induction is a bad inferential strategy. A rational person, says Popper, is one who refuses to use inductive inference; that is, she refuses to extrapolate from past to future, from a finite number of observations to a more general theory, or from a limited number of data-points to a broader pattern. Popper's conviction was that "theories can never be inferred from observation statements, or rationally justified by them. I found Hume's refutation of inductive inference clear and conclusive."[17] Popper therefore set out to show how science could proceed using nothing but deductive reasoning.

Falsificationism

Popper's philosophy of science is founded on an undeniable logical asymmetry. As we have seen, no matter how many individuals you have tested and found to respond positively to *Veritor,* deduction will never tell you that all people respond positively to *Veritor.* On the other hand, if you find just one person who responds badly to *Veritor,* you can conclude—with deductive certainty—that the statement "All people respond positively to *Veritor"* is false. If, as Popper recommends, we need to do science without appeal to inductive reasoning, then while we can never conclude reasonably that scientific generalizations are true, we can conclude that some are false, or so it seems. That is why Popper's view is known as *falsificationism.*

One might think that scientists use a variety of data—from the fossil record, from DNA sequences, from the behavioral and anatomical features of plants and animals—to build a case for a more general claim like "All plants and animals are descendants of a common ancestor." That conception of science, says Popper, is mistaken. Only science founded on induction could aim at the slow accumulation of evidence in favor of particular hypotheses, and Popper regards induction as irrational. Instead, science must proceed by a process of "conjecture and refutation": the scientist begins by formulating a general claim about the nature of the world and then seeks to refute it by gathering data—regarding fossils, DNA, behavior, and anatomy—which, if they go the wrong way, have the potential to show decisively that our general claim about ancestry is false.

This helps us to understand Popper's use of falsification-ism to supply a "criterion of demarcation," which pinpoints the difference between science and what Popper sometimes called "pseudoscience," sometimes "metaphysics." Bona fide science, he says, must be falsifiable. What makes something a genuine piece of science is its potential vulnerability to refutation. Popper was particularly impressed, for example, by the way in which Einstein's relativity theory had laid itself open to the tribunal of experiment. As we will see in more detail a little later, Einstein's theory made explicit predictions for the bending effect that the Sun would have on light arriving at the Earth. It thereby exposed itself to falsification if light turned out not to behave in this way. A properly scientific theory, says Popper, sticks its neck out regarding the sorts of events that it does not permit, hence regarding the sorts of potential pieces of evidence that would lead to the theory being abandoned.

Popper's recipe has considerable intuitive appeal. Freud's theory of the mind is written off as a piece of pseudoscience, because rather than stating in clear ways the sorts of behaviors that would lead to the theory being dropped, Freud offers slippery formulations of his commitments and slippery interpretations of his data. Likewise, the problem with astrology seems to be that its claims are stated in such intolerably vague ways that we cannot judge what it would take for the theory to be shown wrong. Things seem different with astronomy: Newton's theory tells us precisely when to expect the arrival of a comet, and one might think that if things don't turn out that way, so much the worse for Newton's ideas.

The noted physicist Richard Feynman (yet another Nobel laureate) expressed a strikingly similar conception of science—surely influenced by Popper—in a lecture he gave in 1964:[18]

> In general, we look for a new law by the following process. First, we guess it. . . . No, don't laugh, that's really true. Then we compute the consequences of the guess, to see what, if this is right . . . it would imply, and then we compare those computation results to nature, or . . . to experiment or experience. We compare it directly with observations to see if it works.

Feynman continued with a short summary of the falsificationist approach to scientific method:

> If it disagrees with experiment, it's wrong. In that simple statement is the key to science. It doesn't make any difference how beautiful your guess is, it doesn't make a difference how smart you are, who made the guess, or what his name is. If it disagrees with experiment, it's wrong. That's all there is to it.

Gran Sasso

In September 2011 a team of researchers announced that subatomic particles called neutrinos, sent from the CERN facility in Geneva, had been recorded traveling faster than light when their speed was measured at the Gran Sasso facility in Italy.[19] Einstein's special theory of relativity proposes an upper speed limit governing the universe: nothing travels faster than light in a vacuum. Experiment was inconsistent with Einstein's theory. Feynman's summary of the scientific method predicts that in spite of special relativity's beauty, Einstein's name, and his formidable intelligence, the results from Gran Sasso would lead to this esteemed theory being discarded.

This is not what happened. While newspapers lingered for a while on these results, most scientists felt fairly securely that the experimental results were probably flawed. They felt they were flawed partly because of their confidence in the theory those results appeared to contradict. The truth is that scientists do not throw out their theories whenever an experiment appears to contradict them. This attitude is perfectly sensible, because we are often unsure whether experiments have been conducted properly and what their true significance might be. It is perfectly rational to bet on an experiment being flawed, as opposed to putting our money on a well-tested theory being false. This observation causes no trouble at all for the practice of science, but it causes plenty of trouble for Popper's goal of showing how science might proceed without induction.

In the first place, the Gran Sasso experiment shows the limits of the logical asymmetry on which Popper's falsificationism rests. Yes, if our theory tells us nothing can travel faster than light, and if we find something that does travel faster than light, then we know with certainty that the theory is wrong. But just as our judgment of the speed of a car depends on the accuracy of the devices we use to measure it, so we can never simply "observe" how fast a neutrino is traveling, in some self-certifying manner. We must always ask whether the apparatus was working properly, whether we have interpreted our readings correctly, whether our calculations have been appropriate and accurate.

The data, in spite of their name, are not "given" to us in some incontestable manner. Instead, they are the products of hundreds of technical assumptions, any one of which might be challenged. So, if our theory tells us that nothing travels faster than light, and if our experiment indicates that something does travel faster than light, the only thing we are entitled to conclude as a matter of deductive certainty is that somewhere or another at least one mistake has been made. Deduction cannot tell us where that mistake is, and so deduction cannot tell us, by itself, whether our theory is wrong, whether one of our myriad experimental assumptions is wrong, or whether the whole affair is shot through with errors.

Remember Feynman's claim that if a theory "disagrees with experiment, then it's wrong." At Gran Sasso the experiment disagreed with theory, and everyone instead set out to discern what was wrong with the experiment. It is interesting to note elite physicists' reactions a few days after the Gran Sasso result was announced, voiced before any direct evidence emerged for errors in the experimental setup. At this stage, the community had been presented with a result, from an exceptionally well-regarded research group, that appeared to contradict a cherished theory. Martin Rees (the Astronomer Royal, and a recent president of the Royal Society) remarked calmly that "extraordinary claims require extraordinary evidence." The Nobel laureate Steven Weinberg said, "it bothers me that there is plenty of evidence that all sorts of other particles never travel faster than light, while observations of neutrinos are exceptionally difficult."[20] These scientists (and one might cite others) suggested that if forced to bet on whether the established theory or the shocking experimental result was in error, they would put their money on experimental error. These super-luminaries were skeptical of super-luminal velocity.

Rees and Weinberg's sensible skepticism of the Gran Sasso results relies on an inductive inference: it is not available to the strict Popperian, for whom no extrapolation from a solid track record is reasonable. For Rees and Weinberg, the fact that evidence had built up in the past suggesting that other particles do not travel faster than light, and the fact that Einstein's theory itself had held up so well in the face of experimental tests, constituted reasonable grounds for doubting the Gran Sasso result. More generally, when theory and evidence conflict, scientists use inductive inference to help them decide where a mistake has most likely been made. But for the Popperian, such a decision process is irrational.

"Corroboration"

Popper tells us that scientific theories must put themselves up for test. They must stick their necks out and run the gauntlet of experiment. If observation is at odds with theory, then the theory is refuted. A theory may, of course, survive one of these tests, and some theories have survived many rounds of testing. Popper calls these theories highly "corroborated."

Perhaps the most frequently repeated example of this sort of corroboration is Arthur Eddington's experimental test of Einstein's general theory of relativity. As noted earlier, Einstein's theory predicted that light from distant stars would be bent by the gravitational field of the Sun. This bending effect could be observed only during an eclipse, because otherwise the Sun's own brightness would obscure the stars in question. Eddington traveled in 1919 to the island of Principe, off the West African coast, while his colleagues traveled to Sobral in Brazil, in order to be present during a total eclipse of the Sun. Would Einstein's theory be falsified by Eddington's measurements? No: "The results of the expeditions to Sobral and Principe," wrote Eddington and his colleagues, "can leave little doubt that a deflection of light takes place in the neighbourhood of the Sun and that it is of the amount demanded by Einstein's generalised theory of relativity, as attributable to the Sun's gravitational field."[21]

Eddington's results are typically thought of these days as providing strong evidence in favor of Einstein's theory. But when Popper says that a theory is highly "corroborated," he does not mean that the theory is likely to be correct. "Corroboration" is merely a statement of a theory's past success, and since, for Popper, past success provides no guide whatsoever for future prospects—to think it did would involve a form of inductive inference—this also means we have no reason to think a highly corroborated theory is likely to pass the next test thrown at it.[22]

There is a sense in which, for Popper, our credence in a scientific hypothesis should be unaffected by whether the theory in question has just been plucked from thin air, or whether instead it has a long and distinguished track record of remarkable success in the face of searching experiment. Since corroboration can bear no weight for Popper, this also makes it hard to see how, on Popper's view, scientists like Rees or Weinberg could ever be justified in thinking that because Einstein's ideas have held up so well in the face of severe tests, our suspicions should probably lie with the manner in which the equipment at Gran Sasso was set up.

Theory and Observation

What is the status of the pieces of data, or reports of observations, that the falsificationist thinks scientists can use to reject general theories? Popper insists, with good reason, that observation is "theory-laden." Roughly speaking, this means that apparently neutral statements about observational data are invariably shot through with assumptions about scientific theory. For example, a claim like "We observed a neutrino travel in excess of the speed of light" can be made only when a vast amount of knowledge is presupposed about how neutrinos behave, how they can be detected, and how our instruments work. By itself, this dependence of observation on theory is unproblematic: indeed, if scientific observation were not enabled by theory, then the ability of scientists to probe the inner workings of the universe could not make progress. But the "theory-ladenness of observation," as philosophers like to call it, leads to special problems for Popper.

Popper's rejection of induction means he denies that limited numbers of observations can ever provide support for general theoretical claims. But he also recognizes that statements about what has been observed—what scientists would usually call their data—rely on general theoretical claims as well. In fact, Popper takes the view that *all* "observation statements" are laden with theory—not just exotic claims about how fast a neutrino has traveled but apparently more banal claims about whether a piece of litmus paper turned blue, whether a Geiger counter registered a click, and so forth. Since the data presuppose theory, Popper concludes that observation statements are no less conjectural—hence no less provisional—than the theories they are supposed to falsify.

Popper's deductive method is far less powerful than we might initially think. On the face of things, Popper offers us the consoling thought that even if we can never conclude reasonably that a theory is likely to be true, we can at least conclude that some theories are false. But showing that a theory is false requires that we have justified confidence in the observations that we use to refute the theory in question. If observations themselves are mere conjectures that draw on general theories, and if those general theories cannot be supported by induction, then this confidence can never be had. What scientists can do, in Popper's scheme, is to show that one set of statements—general ones, about how things work—are in logical tension with another set of statements—specific ones, about particular events. Science cannot give us any confidence about which, if any, of these statements are likely to be correct. Science cannot do this, so long as it shuns inductive inference.

Piles in a Swamp

When observation and theory clash, how does Popper think scientists are supposed to decide whether to discard theory (on the grounds that the observations in tension with it are to be trusted) or observation (on the grounds that it has been generated through dubious experiment)? Popper's stance on the status of observation statements is striking:

> Science does not rest upon solid bedrock. The bold structure of its theories rises, as it were, above a swamp. It is like a building erected on piles. The piles are driven down from above into the swamp, but not down to any natural or given "base": and if we stop driving the piles deeper, it is not because we have reached firm ground. We simply stop when we are satisfied that the piles are firm enough to carry the structure, at least for the time being.[23]

The thought that science "does not rest upon solid bedrock" might be comforting to those humble scientists who rightly stress the fallibility of their work. Only a fool would claim cast-iron certainty for a piece of experimental data. But Popper's piles give him discomfort. Sink piles into a swamp, and they have something to grip on. It is possible to build there. But what weight can observation carry, once induction has been rejected?

Popper thinks that we can use a certain class of observation statements—namely, the ones we "decide to accept"—as the basis for the falsification of theories. These are the statements the scientific community views as uncontroversial. Popper calls them "basic statements." But one hopes that science is built on more than mere group agreement. It is important that scientists' judgments about acceptable observation statements are shared because those judgments are also reasonable, or reliable. On the matter of the reliability of observation, Popper has nothing to say:

> The basic statements at which we stop, which we decide to accept as satisfactory, and as sufficiently tested, have admittedly the character of *dogmas,* but only in so far as we may desist from justifying them by further arguments (or by further tests). But this kind of dogmatism is innocuous since, should the need arise, these statements can easily be tested further. I admit that this too makes the chain of deduction in principle infinite. But this kind of "*infinite regress*" is also innocuous since in our theory there is no question of trying to prove any statements by means of it.[24]

Popper tells us that, in practice, scientists can decide whether it is theory or observation that is at fault, because the community simply accepts, by common convention, that a certain class of observation statements will be viewed as unproblematic. If a theory disagrees with these statements, then so much the worse for that theory. But group endorsement might arise from all sorts of pathological sources. What answer can Popper give to the skeptic who says that the data-points science aims to systematize are merely the product of collective fantasy or collective conspiracy?

The strict deductivist cannot justify the decision to regard these data as "satisfactory, as sufficiently tested," by appeal to their track record, because the thought that these claims have held up so well that they are likely to be true is a piece of inductive inference. The deductivist can, of course, point to the possibility of evaluating these statements, by subjecting them to further test. Hence they are not pure dogma. But these tests, too, involve seeing how our supposed observations tally with other forms of equally conjectural data.

And so we ask our question again: What makes any of these conjectures anything more than collective confabulation? Popper thinks the regress innocuous because proof is not the aim of science. This gives the impression that we can settle for something short of proof: reasonable grounds, or a decent justification for our observation claims. But on Popper's view we have no reason for thinking that observation statements are reliable, or trustworthy. Once we deny ourselves induction, we lose any chance that our theories might grip onto reality. Popper's scientific edifice is not a building erected on piles in a swamp; it is a castle in the air.

Popper and Popularity

When we think of the gilded Knights of the British Empire and Fellows of the Royal Society who have queued up to endorse Popper's image of science, it will perhaps be a surprise to learn that, on Popper's view, we have no reason whatsoever to think that our best scientific theories are true, close to the truth, or even likely to be close to the truth. These worries about Popper's system are not new: several generations of undergraduate students have trotted out similar lines of attack. Why, then, does Popper continue to be held in such high esteem by so many scientists?

Part of the reason, of course, is reciprocity: Popper himself had unwavering respect for the work of the sciences, and scientists feel they should return the favor. I also suspect that when scientists read Popper, they come away with a watered-down, more palatable form of Popperianism, one that overlooks Popper's strict skepticism of inductive inference. Popper says science does not deal in certainty. He is right about this. Scientists are keen to stress that their theories are never held dogmatically, that they are always open to challenge, that even long-held theories might fall prey to uncomfortable facts, and that scientific data, just as much as scientific theories, are hard to attain, and potentially revisable. But note how far this sensible form of fallibilism—"we might have got it wrong"—is from Popper's anti-inductivism—"there is no reason to think we have got it right." It is the difference between acknowledging that Usain Bolt might stumble and lose and arguing that there is no reason to think Bolt will go faster than anyone else who happens to be running.

Popper also stresses that, in designing experiments, scientists are not simply looking to collect facts that their theories can account for. Again, he is right about this. Scientists praise Popper for understanding that they are trying to ask probing questions of nature. An experiment should be designed so that if its results go one way the theory it tests will be in trouble, whereas if they go the other way the theory receives an evidential boost. Scientists take Popper's insistence on falsifiability to be a means of stressing the importance of demanding tests. But I suspect few scientists would agree with Popper that even when many of these tests have been passed, we have no reason for placing any confidence in the theory; and I suspect even fewer would accept his view that the standing of both theory and evidence is ultimately a matter of collective convention. Popper's philosophy of science is not the mild view that science is a fallible enterprise, which seeks demanding tests for its theories.

Demarcation Revisited

It is possible to isolate an eviscerated and attractive Popperianism that does away with Popper's own strict rejection of induction, stressing instead the important themes of testability and fallibility. What are the prospects for using this sort of mild falsificationism for the purposes of demarcation? Is a genuinely scientific theory one that is testable?

For a theory to be testable, it needs to make predictions. No theory—not even an intuitively "scientific" one that we think should fall on the good side of the demarcation line—makes predictions all by itself. Newton's laws of motion, taken on their own, do not tell us where we will observe objects. Darwin's principle of natural selection does not tell us all by itself what sorts of organisms will exist. Instead, these theories make predictions only when they are supplemented with a whole catalog of additional assumptions.

If one adds to Newton's laws a rich set of claims about where objects are located, how massive they are, and so forth, then we can use those laws to make predictions about these objects' later locations. If one adds to Darwin's principle of natural selection an even richer set of claims about genetic mutation rates, developmental processes, typical interactions between species members, and so forth, then that principle, too, can tell us something about how a species will change over time. So we cannot fault intelligent-design theory, or astrology, on the grounds that they make no concrete predictions, for no theory makes predictions when considered in isolation.

Moreover, just like Newton's laws or Darwin's principle of natural selection, these theories can be supplemented with additional assumptions so that they do make specific predictions: in other words, astrology and intelligent-design theory can become falsifiable. There is nothing to stop an astrologer foretelling in rather specific terms that Cancerians like me will have a nasty accident next Tuesday; there is nothing to stop an intelligent-design theorist from predicting that, since God is wise and beneficent, human anatomy in general will turn out to be well designed. But what will the astrologer say if everything seems to go fine for me next Tuesday? What will the intelligent-design theorist say if an anatomist points out the apparently perverse layout of the male urinary system, which requires the urethra to pass inside the prostate gland, causing misery for men when the prostate becomes enlarged and the urethra becomes constricted? If we want to use a Popperian criterion to determine the scientific status of theories, we need to focus on how the theorists responsible for them handle failed predictions. Unfortunately, there doesn't seem to be any clear recipe that will tell us what sort of response is "scientific," and what sort of response is "unscientific."

We do not want to say that a theory is scientific only if the theorists who put it forward are prepared to reject it the moment its predictions appear to be contradicted by experiment. It is perfectly reasonable for a theorist to dig in and say that, while the experiment might seem to be bad news for the theory, she believes fault to lie with the experimental setup itself. That is exactly how the scientific elite responded to the apparent demonstration of faster-than-light neutrinos at Gran Sasso. But if particle physicists are allowed to evade refutation by suggesting that the blame for a failed prediction does not lie with their theories, but lies instead with other factors external to those theories, then what is to stop the astrologist, or the intelligent-design theorist, from pointing the finger at something other than the view that our lives are influenced by the stars, or something other than the view that organic traits are the products of conscious design, when I fail to have an accident on a Tuesday, or when my prostate swells to constrict my urethra? Cannot they, too, offload the blame for failed prediction on an error of calculation, or a hidden assumption, or a misunderstanding of the theory itself? What, precisely, is the difference between an intelligent-design theorist telling us that we cannot fathom God's peculiar intentions for my urinary anatomy and a physicist insisting that the apparatus at Gran Sasso must have been malfunctioning in some as-yet-undetermined way? Don't all of these theorists use similar tactics to preserve their theories from refutation?

The obvious response to all of this is to say that the difference between the scientific and the nonscientific attitudes is a matter of how shameless one is when it comes to persistently delaying the rejection of a theory, in favor of rejigging one's ancillary assumptions. A view of this broad variety—greatly elaborated and backed up by historical examples—was defended by Popper's admirer and LSE colleague Imre Lakatos.

Newton's laws were used to predict the orbit of Uranus. Uranus was instead found to take a course different from the predicted one. Astronomers refused to reject the Newtonian framework, suggesting instead that perhaps an unknown planet was pulling Uranus off course. Such a move would seem desperate—a blatant case of evading the tribunal of experiment—except that the new planet Neptune was subsequently discovered in just the position required to disturb Uranus's orbit. And when our particle physicists suggested that something might be awry in the Gran Sasso experiment, their bet also paid off in the end: it was subsequently confirmed that a fast-running clock and a faulty connection had combined to produce a mistaken calculation for the journey time of the neutrinos.[25]

The examples of Neptune and Gran Sasso are vindications of a refusal to relinquish a good theory in the face of problematic evidence. But note how difficult it is to turn these anecdotes into a hard-and-fast set of rules regarding scientific status. Is a scientist being suitably tenacious in the face of experimental adversity, developing a masterful theory whose confirming data is just around the corner? Or is he just being pig-headed in response to a manifest lack of evidence in favor of his views?

Looking back, it is tempting to credit Darwin, for example, with a kind of prescient knowledge of the merits of his theory. His claim that the diverse species of plants and animals are all descended by gradual steps from a small number of common ancestors has the implication that some time in the past there must have been species whose anatomy and physiology fill in the gaps between the distinct forms we see today. Darwin was not able to point to such intermediate forms. He argued that his inability to produce them did not constitute a problem for his theory, but was instead a symptom of the rarity with which fossils are preserved.[26] We can give Darwin credit *in retrospect*, because in the intervening years we have discovered many "missing links," each of which adds further support to Darwin's view of common descent. But how are we to apply this sort of criterion *prospectively*, if what we want to do is sort the scientific wheat from the pseudoscientific chaff right now?

"The Inquiring Mindset"

Popper is of little help if we want a practical, prospective criterion of demarcation. In spite of everything that we read about the importance of the "scientific method," it remains unclear what that method is. The basic mathematical tools of statistical inference form a fairly constant part of the scientist's toolkit. There are also, of course, plenty of scientific *methods*: there are techniques of observation and analysis specific to individual sciences. We can use randomized controlled trials for understanding the efficacy of medicines, we can use X-ray crystallography for understanding the structure of molecules. But when we try to pinpoint a recipe for inquiry that all successful sciences have in common, we run into trouble.

Yet another Nobel laureate, Sir Harry Kroto, suggested in *The Guardian* a few years ago that we may have to settle for a loose account: "The scientific method is based on what I prefer to call the inquiring mindset."[27] The scientist approaches nature in a spirit of curiosity; she asks honest questions of nature. She proposes a hypothesis and seeks out evidence, often through a well-designed experiment, that will adjudicate on the truth of that hypothesis. But while this account does indeed help us to explain what makes science an admirable activity, it does not isolate a method that distinguishes the sciences from other branches of inquiry. Historians, too, can propose bold hypotheses, before delving into a historical archive in the spirit of honest inquiry. The same goes for other researchers in the humanities.

Kroto added to his very capacious remark on "the inquiring mindset" that this favored attitude "includes all areas of human thoughtful activity that categorically eschew 'belief,' the enemy of rationality. This mindset is a nebulous mixture of doubt, questioning, observation, experiment and, above all, curiosity, which small children possess in spades."[28] Kroto is right, of course, to stress that the sciences, as traditionally understood, do not have a monopoly on critical inquiry. But his doubts over the value of "belief" overlook the positive role of stubborn dogma. As we have seen, good scientists do not reject a theory the moment it fails to line up with experimental data. Instead, they frequently throw the blame for failure onto an unknown fault with their equipment, an unreliable observation, or a whole mistaken tradition that has led to a misunderstanding of what the apparent "evidence" amounts to. These sorts of tactics—which may look for many years like head-in-the-sand obfuscation, and which are regarded only in the light of later evidence as the foresight of genius—are often productive.

The value of blind conviction in producing valuable scientific results is one of the central themes of Paul Feyerabend's notorious book *Against Method:*

> Newton's theory of gravitation was beset, from the very beginning, by difficulties serious enough to provide material for refutation. Even quite recently and in the non-relativistic domain it could be said that there "exist numerous discrepancies between observation and theory." Bohr's atomic model was introduced, and retained, in the face of precise and unshakeable contrary evidence. The special theory of relativity was retained despite Kaufmann's unambiguous results of 1906.[29]

Feyerabend is alluding all too briskly to a series of theories—due to Isaac Newton, Niels Bohr, and Albert Einstein—which we now take to be triumphs of scientific inquiry, and which were kept alive in infancy in spite of the problems they faced. Newton, for example, was not able to explain why the solar system should be a regular system at all. Why wasn't it thrown into chaos by the mutual gravitational attractions of planets and comets? Bohr proposed that the atom itself is similar in structure to the solar system, with electrons orbiting a central nucleus. His initial model was unable to account for data concerning the behavior of hydrogen when it emits energy—particularly the so-called Pickering-Fowler ultraviolet series—that was known before Bohr's model was put forward, and which was explained by a rival theory. Walter Kaufmann's experiment of 1906, which aimed to determine whether electrons were rigid spheres or whether they could instead be deformed (as Einstein's theory seemed to entail), was widely thought at the time to have produced a result at odds with Einstein's theory of the electron.

Feyerabend's language is inflammatory, but his underlying argument is a reasonable one. In claiming that Newton's views could have been *refuted*, he implies that they could have been proven false. In claiming that the contrary evidence against Bohr was *unshakeable*, he implies that this theory, too, was known to be false at the moment it was introduced. We do not need to go this far to see that Newton's theory, and the others he mentions, were borne into hostile evidential environments. It took time, for example, for Bohr to develop a model of the atom that could account for the problematic Pickering-Fowler series. Feyerabend is surely right in saying that if scientists didn't sometimes stick resolutely to their theories in spite of abundant problems that seem—perhaps mistakenly—to undermine them, then the scientists in question would never be able to develop both mature theory and a properly interpreted body of evidence, of the sort that future generations take to be indicative of a visionary scientific achievement. The scientific mind is often open, creative, and sensitive to evidential detail. But sometimes scientists, like horses, progress best when their blinders are on.

Further Reading

On Popper's life, see his autobiography:
Karl Popper, *Unended Quest: An Intellectual Autobiography* (London: Routledge, 1992).

Popper's own writings are highly accessible, especially the following:
Karl Popper, *Conjectures and Refutations: The Growth of Scientific Knowledge* (London: Routledge, 1963).
Karl Popper, *The Logic of Scientific Discovery* (London: Routledge, 1992).

Most introductions to the philosophy of science include discussions of Popper's work. A lively (and uncharitable) critique can be found in:
David C. Stove, *Popper and After: Four Modern Irrationalists* (Oxford: Pergamon, 1982).

Meanwhile, a far more sympathetic assessment of Popper's work is provided in:
David Miller, *Critical Rationalism: A Restatement and Defence* (Chicago: Open Court, 1994).

For a sophisticated form of Popperianism that aims to bring Popper's basic views into alignment with the history of science, see:
Imre Lakatos, *The Methodology of Scientific Research Programmes* (Cambridge: Cambridge University Press, 1980).

Chapter Two

Is *That* Science?

The Diversity of Knowledge

The conclusion of the last chapter might have left readers feeling uneasy. If we cannot appeal to Popper's falsificationism to give us a criterion of demarcation between science and pseudoscience, then what can we say if, like Popper, we have a nagging sense that not all realms of inquiry are equally respectable? How can we assess the standing of disciplines that are sometimes alleged to be spurious examples of science, such as economics, intelligent-design theory, and homeopathy? Are we reduced to an admission that, in the realm of science, anything goes?

Fortunately, in spite of our rejection of the Popperian philosophy, we still have plenty of resources left to permit a critical evaluation of these contested areas of investigation. Rather than asking whether, in general terms, economics, intelligent-design theory, and homeopathy are "scientific," we instead need to ask more specific questions—about the roles of idealization in science, about evidence, and even about the nature of placebos—if we are to put our finger on what is troubling about these different projects.

Sometimes little hangs on the general question "Is it science or not?" Should we say, for example, that history is a science? Few people would think of it in this way, and yet history shares with the natural sciences an attitude of critical inquiry, sometimes peppered with the sort of productive dogmatism that allows a controversial theory to be developed and to orient itself to bodies of properly interpreted evidence. Historians gather data from a variety of sources in order to test their conjectures, and while they rarely perform experiments, we must remember that the same can be said of some canonical sciences. Astronomy, for example, is more often in the business of making observations than setting up controlled investigations in the laboratory.

One might try to justify the verdict that history is not a science by pointing out that historians typically focus on coming to a fine-grained understanding of particular, contingent events rather than aiming to construct general laws; but this, too, is arguably a feature of a science like evolutionary biology, which accounts for the idiosyncratic makeup of particular species, and which has its own historical focus on understanding branching patterns of descent from common ancestors.

Often history is free of mathematical analysis, but some forms of history—especially economic and demographic history—are rich in quantitative and statistical detail. History deals with human action and human decision, but so, too, do psychology and anthropology. In short, the sciences are diverse enough in their own practice that it is largely a matter of taste as to whether history is classified with them. We do not typically call history a science, but it would not be an outrage if we did: in the German language the term *Wissenschaft* is used to denote any disciplined approach to the generation of knowledge, and it thereby encompasses subjects that English speakers would intuitively classify as sciences and as humanities. Sometimes, then, questions of demarcation do not matter much. Sometimes, however, they carry considerable weight.

Economics and the Ideal

Alfred Nobel's will of 1895 made provision for five prizes in his name, to be awarded in the areas of physics, chemistry, physiology or medicine, literature, and peace. What about the Nobel Prize in economics? That prize was a latecomer, endowed "in memory of Alfred Nobel" by the Sveriges Riksbank in 1968, for work in the area of "economic science."[1] But merely calling something a science does not make it so, as the examples of "creation science" and "Christian Science" remind us. Is economics a true science, or is the generosity of the Sveriges Riksbank best understood as an effort to see that some scientific sparkle rubs off from physics, chemistry, and physiology onto a field that does not merit it?

The diversity of scientific practice, combined with the diversity of approaches to economics, makes this a difficult question to answer in any straightforward way. Some styles of economics involve an experimental rigor and curiosity that allies them closely with work in experimental psychology. Some economists, for example, are interested in understanding how real people make real decisions, and they place people in laboratories in order to find out. Daniel Kahneman's 2002 prize in memory of Alfred Nobel was awarded for experimental work of this sort. Kahneman's research (much of it done in collaboration with Amos Tversky) has aimed to demonstrate the ways in which people think, especially the rules of thumb they use when making judgments about uncertain events.[2] A few researchers have gone so far as to investigate how economic decision-making differs from one culture to another.[3] This sort of work has as good a claim as any to the status of science.

The economist Amartya Sen won the Sveriges Riksbank prize in 1998, and his work, too, can hardly be accused of paying insufficient attention to the details of how things are. One of Sen's most famous pieces of work concerns the causes of famines.[4] It might seem obvious that famines are caused by a general decline in the availability of food. Sen argues, with painstaking attention to empirical data, that this is not the best explanation: on many occasions, famines can occur with no decline in food availability. Instead, the question to ask is why, in a famine, some people are unable to get their hands on the food that is available. Sen's answer, which points to the ways in which people acquire power to amass resources, suggests a variety of practical ways to reduce the incidence of starvation. It is hard to see why we should not count this work, alongside that of Kahneman, as bona fide science.

In contrast to these empirically rich forms of economic inquiry, much work in neoclassical economics is concerned with the largely theoretical analysis of how markets would work if they were populated with individuals endowed with perfect rationality—in other words, creatures of fantasy. We might be tempted to classify these areas of economics as science fiction. Alternatively, we might think that this brand of economics tells us not how the world is but how the world ought to be, if only people would think straight. Both reactions suggest a gulf between neoclassical economics and the typical practice of science. Both reactions are too hasty.

Economics is not alone in its use of simplification and idealization.[5] Simple physics can show us how far a cannonball would travel if it were subject only to the force of gravity and the force imparted by the ignition of the gunpowder. Of course, no real cannonball is like this: a real cannonball is subject to other forces, like wind-resistance. It does not follow that our simplified analysis of the ball's trajectory is without value. First, it helps us understand something about the cannonball's basic tendencies, which may sometimes be impeded by other forces that are too complex for us to take into account. Second, if we can measure how far a real cannonball travels, and if we compare this with our analysis of how far the cannonball would travel if it were affected only by gravity and its initial accelerative force, then we have clues about the nature of the other forces that must have prevented the real cannonball from traveling the distance predicted by our simplified calculation. In this way, unrealistic idealizations help us to understand more complex real-world events.

Physics is not the only science to make use of idealization. Significant amounts of research in biology explore, in mathematically sophisticated ways, the action of idealized evolutionary forces on idealized organisms. Evolutionary geneticists often construct simplified theoretical models, which assume that populations of organisms are infinite in size, or that genes interact with each other only in the simplest of ways, or that all organisms reproduce at the same time, or that natural selection has no effect. Evidently none of these things is true for populations of real plants or animals in the wild. Again, though, if we compare the real behavior of a population with a simplified analysis that tells us how a population would behave under the assumption that there is no natural selection at work, we can estimate whether natural selection is in fact at work, and if so how significant its effect has been.[6]

Might economists defend their own practices of idealization in a similar way? Could they argue that, just like many other scientists, they are simply investigating a simplified set of basic tendencies for human behavior, as a preliminary to adding further complicating details?[7] Perhaps, but idealization of this sort ultimately needs to pay its way in the currency of empirical testing. To see why, we need to think again about simplification in physics.

We cannot invent any set of assumptions we like about how a cannonball will behave in simplified circumstances where gravity alone acts, and wind-resistance does not. Unconstrained invention of these supposed "tendencies" will not help us to understand what additional real-world forces stop the real cannonball from traveling as far as our simple model predicts. On the contrary, depending on how our caprice leads us to calculate the simplified behavior of the ball, a real cannonball might end up traveling precisely as far as our simple model predicts, or twice as far, or half as far.

We will then be led to a variety of wholly misleading claims about what additional real-world forces, not included in our simple model, must have produced these divergences. Our assumptions about the cannonball's tendencies need to reflect the realities of its behavior when unimpeded by wind. In other words, even when we introduce simplifications, these simplifications must be disciplined by experiment. That is why physicists are careful to examine the behavior of objects in the controlled circumstances of the laboratory, where complicating factors can be reduced to a minimum.[8] There is no sin in idealization, but economists cannot use idealization as an excuse to keep their hands clean of the dirt of real human thought and behavior.

As a final thought on economics, we should recall that those critics who faulted neoclassical economics for its failure to predict the financial shocks of 2008 are perhaps best understood as complaining, not that economics is unlike true science, but rather that it is too much like true science. Physicists, as we have already seen, often deal in claims about the basic tendencies of objects. They excel at predicting what happens in the sanitized conditions of the laboratory, where circumstances can be kept as simple as possible. For that reason it is not physicists, but engineers, whom we approach when we want a serviceable structure that can survive the vicissitudes of the world outside the lab. Physicists do not build good bridges. If economists are to offer practical advice to government, then we do not want economics to be like fundamental science; we want it to be like engineering.[9]

Evidence and Intelligent Design

Open a textbook of evolutionary biology, and you will find a large body of mathematically detailed principles describing how populations are likely to change under the influence of various evolutionary forces, supplemented with copious experimental data regarding the processes by which genetic mutation occurs and the ways in which species members interact with each other and with their surroundings.[10] You will find that all of this information is stitched together in ways that aim at detailed specifications of the conditions under which these processes can—and cannot—result in the production of new species ("speciation") and the production of adaptation to species' environments. You will find studies of natural selection at work in the wild, and, of course, you will find debate about the relative importance of different evolutionary processes, disciplined by careful experiment.[11]

The theory of intelligent design—a theory defended by the likes of the American mathematician and philosopher William Dembski and the American biochemist Michael Behe—purports to be a scientific theory that explains at least some instances of organic adaptation in terms of design by an intelligent agent.[12] For example, Michael Behe takes the view that the *flagellum*—a twirling whip-like filament, attached to a structure rather like a rotary motor, which propels certain types of bacteria through a liquid medium—is too delicately organized for it to have been produced by natural selection. He argues that it has been designed by a being endowed with intelligence. Intelligent-design theorists have held back from saying that organic adaptations such as the flagellum have been produced either by some God or another, or by the Christian God in particular. Typically, they have suggested that their evidence points to some sort of overseeing intelligence, but they have resisted saying much more about the deeper nature of that intelligence.[13]

How does the argument for intelligent design, and against natural selection, work?[14] Behe, in his book *Darwin's Black Box*, claimed that the bacterial flagellum (along with various other traits) exhibits a property he calls "irreducible complexity."[15] This means that if parts of the flagellum were removed, or altered, then rather than impairing the performance of the flagellum a little, we would instead find a structure that makes no contribution whatsoever to the survival and reproduction of the bacterium. The overall operation of the flagellum, says Behe, is so finely orchestrated that any disruption to any one of its parts would be catastrophic for any sort of valuable biological function.

Behe takes the view that natural selection builds complex traits by gradual improvement from simple beginnings. He consequently argues that if we really were to find a structure whose overall function would be ruined entirely by the removal or alteration of a single part—if, that is, we found something "irreducibly complex"—then such a structure could not possibly have been produced by this process of gradual accretion. Irreducible complexity, Behe thinks, is perfectly diagnostic of the explanatory impotence of natural selection.

The first thing to note in response to Behe is that the flagellum is most likely *not* irreducibly complex. Research suggests that a partial flagellum is indeed biologically useful: instead of producing rotary locomotion, it allows protein toxins to be injected into other cells.[16] But even if further research forced us to conclude that the flagellum is irreducibly complex, Behe is wrong to think that irreducible complexity is incompatible with explanation by natural selection. We must consider the possibility that natural selection first builds a rather ramshackle structure, by gradual steps over a long period of time. Such a structure may have plenty of scope for its elements to be removed or altered, and for overall function to be damaged only a little. Selection might then gradually remove useless or redundant elements—in the interests of overall economy—so that what we are left with in the end is a structure that is "irreducibly complex" in the sense that any further removal of, or tinkering with, its parts would ruin functioning entirely, in spite of the fact that selection is perfectly capable of accounting for its genesis in a gradual manner. Of course, it would be reasonable for Behe to complain that we have no evidence that this is, in fact, how the flagellum was produced. He might say that our hypothesis for slow refinement from simple beginnings is entirely speculative. He would be right about that, but Behe's case for intelligent design rests on the claim that there is no *possible* means for selection to account for the flagellum, and a speculative sketch is enough to refute such a strong claim of impossibility.

Suppose we follow Behe even further, and agree that there are some biological structures that cannot be explained by natural selection. This would not tell us anything at all about what does explain them: it would show only that there are some things we do not understand. What case is made, then, for thinking that intelligent design is what explains the flagellum? The answer is not at all obvious, because it is unclear how the existence of an intelligent designer is supposed to account for the flagellum's properties. Imagine I tell you: "There are intelligent designers on Mars." This gives you no insight at all into the kinds of designed structures one might expect to observe there unless I also give you information about how intelligent these designers are, what size they are, how lazy they are, how cooperative they are, their economic priorities, the sorts of raw materials they have access to, and so forth.

To have a satisfying explanation for the bacterial flagellum, then, we need our design theory to spell out in some detail the tools and capacities the designers are supposed to have, the design brief they are supposed to be following, the manner in which they go about refining and enacting their designs, the building materials they have access to.[17] These details—about how the processes of organic construction are supposed to work—are supplied in abundance by evolutionists, and they are not supplied at all by intelligent-design theorists.

What is more, biologists do not merely hazard the nature of evolutionary processes and have done with it: instead, having put together a set of hypotheses for genetic mutation rates, the strength of selection, and so forth, they test these assumptions directly via experiment. Intelligent-design theorists, on the other hand, neither spell out in detail what the situation of their alleged designers is supposed to be like nor move on to test their conjectures experimentally. That is why evolutionary explanations of change in the organic world have strong levels of evidential support, whereas the level of evidential support enjoyed by intelligent-design theory is risible.[18]

In one sense, the theory of intelligent design competes directly with evolutionary biology: intelligent-design theorists and evolutionary biologists disagree about the processes that have shaped structures like the flagellum. But if intelligent-design theory is to be judged as a *serious* competitor—that is, as a theory with strong credentials—then it, too, would need to have accumulated a rich body of theory regarding speciation and adaptation, all informed by painstaking experiment. It, too, would have given us textbooks giving detailed accounts of how exactly the alleged design process works. Under what circumstances are design processes overwhelmed by other forces, and when do they dominate them? What is the nature of the proposed designer, and in what ways is that designer constrained to act? When confronted with conflicting design specifications, how does the designer typically undertake trade-offs? We would expect thorough answers to all of these questions. Of course, we find no such answers. Instead, the intelligent-design theorist treats us to a catalog—often a very big one—of organic structures whose complexity, it is alleged, is such that it could not possibly be explained by natural selection. These are not adequate grounds for the theory to be taken seriously.

Is it possible for me to side with Paul Feyerabend, as I did in the last chapter, on the need for scientific theories to mature until they reach a point at which they can be brought into contact with problematic counter-evidence, while also dismissing intelligent design on the grounds that it has no appreciable evidential backing? In Chapter 1 I suggested that we cannot tell in advance which theories will end up being well supported by data, and for this reason it would be methodologically disastrous to smother at birth a theory destined for greatness on the grounds that in this early period of development there is no evidence in its favor. Mindful of the fallibility of science, we should not rule out the thought that, at some time in the future, a group of investigators might put together a theory that hypothesizes in some detail how, for example, a group of intelligent aliens have been able to intervene in the makeup of particular earthly plant and animal species.

Such a theory might later receive direct support, if our future theorists were to observe the modus operandi of these intelligent agents. It is not impossible that one day we might find the intellectual property rights of these master artists asserted in the DNA of diverse species, or even that we might catch their emissaries—maybe tiny intelligent robots—in the act of fashioning new adaptations in response to changed environments. But these are merely possibilities. As things stand, intelligent-design theory certainly does not deserve to be taught in schools as a bona fide alternative to evolutionary theory, and neither do my silly speculations about intelligent alien bio-engineers.

Homeopathy and the Nature of Placebo

Homeopathic medicines are created, for the most part, by taking small quantities of plants and minerals that are repeatedly diluted in water, shaken vigorously, and diluted again before being added to sugar pills. The substances in question are selected using the hypothesis that "like cures like": if a substance would harm you in high concentration, the same substance taken in an exceptionally weak dosage will combat those very symptoms, or so the story goes.

Homeopathic remedies are popular: the British Homeopathic Association estimates that they are used by over 200 million people around the world.[19] Nonetheless, many commentators have been suspicious of homeopathic medicine, and it is not hard to see why. Smallpox was initially controlled by deliberately placing infectious material from smallpox pustules into a scratch on a healthy person's skin—a process known as inoculation. So the "like cures like" principle is not obviously absurd. But homeopathic treatments are diluted and rediluted so many times that we should often expect there to be *no* molecules of the supposedly therapeutic substance present. This means that a homeopathic treatment is sometimes nothing more than a sugar pill with a lively backstory. Is homeopathy founded on good science? Or is it pseudoscientific quackery?

It is sometimes thought that the "gold standard" for testing the efficacy of a medical intervention of any kind is the placebo-controlled randomized clinical trial. In the case of a new drug, for example, the idea is that the best way to assess its merits involves taking a large number of people and dividing them into two groups by random allocation. Half are given the new drug, half are given sugar pills. Since we know that the act of swallowing a pill under doctor's orders—even when the pill is merely made of sugar—can make people feel better, this method allows us to evaluate the additional boost to health delivered by the new drug, beyond that which we can credit to the placebo effect.

Many theorists have questioned the wisdom of insisting on placebo-controlled trials.[20] After all, there are plenty of ways to see whether a medicine is beneficial. Instead of asking whether a new drug is better than placebo, you can ask whether it is better than the standard treatment used for the condition in question. This will also tell you whether the new drug gives a therapeutic boost beyond that delivered by the placebo effect (for the efficacy of the standard treatment will also have a placebo element). Moreover, you do not simply learn whether the new drug works, you also learn whether it works better than what we already have. That second piece of knowledge is especially valuable for doctors deciding which medicines to give to their patients. What is more, disease-sufferers recruited to the drug trial do not need to interrupt their therapeutic regime by moving onto sugar pills for a temporary period: they can continue to use the standard treatment.

Trials of new drugs compared with standard treatment can be more expensive than placebo-controlled trials because the numbers of people involved usually need to be higher in order to get a statistically meaningful signal. It is also much easier to produce a new drug that is better than placebo than it is to produce a new drug that is better than the standard treatment. For those reasons critics have sometimes argued that the custom of testing new drugs against placebo, rather than against standard treatment, has become widespread because it is in the interests of large pharmaceutical companies.[21]

Homeopathy is just one instance of the broader category of complementary and alternative medicine. Other techniques that fall into that category include acupuncture, aromatherapy, and herbalism. Proponents of alternative and complementary therapies occasionally argue that the requirement to test treatments against placebo is especially unkind to them. Suppose you become convinced, on the basis of anecdotal experience, that massage therapy is an effective treatment for neck pain. How are we to test your claim using a placebo-controlled trial? In the case of a new drug, the creation of a placebo to match it involves designing something that seems to the unwitting patient just like the real drug, but which lacks the active ingredient. That is why we use a colored sugar pill. But what would it mean to give someone placebo massage? We would need to create an intervention that seems to the patient just like "real" massage, even though it lacks whatever the "active" element in massage might be. It is hard to see how we can create the illusion that someone is genuinely receiving massage without giving him a massage. The resulting confusion explains why, when attempts have been made at placebo-controlled trials of massage therapy, "placebo massage" has sometimes been understood as massage "without pressure," and sometimes it has simply involved playing music to people without touching them at all.[22]

Let us return to the particular problems presented by homeopathy, rather than the problems facing the assessment of complementary and alternative therapies in general. We might think it fairly straightforward to subject homeopathic medicines to placebo-controlled trials. We can simply give patients sugar pills that have no history of contact with the supposedly active substances of homeopathic therapies. The British Homeopathic Association raises a rather different problem with the idea that we can use controlled trials to evaluate homeopathy. In mainstream medicine we are familiar with the idea that depression is treated with Prozac, or that high cholesterol is treated with Lipitor. In other words, there is an assumption built into these trials that a particular kind of intervention is to be assessed for a particular kind of condition.

But what if a doctor does not assign a type of medicine to a type of disease condition, but instead she attempts to evaluate the overall situation of the person arriving at the clinic? Suppose she claims that symptoms, biography, and lifestyle all interact in complex ways, and she offers her patient a bespoke blend of medicines, advice on how to exercise, what to eat, and so forth, all in the conviction that these act in a synergistic, integrated way on the person as a whole? It will then be difficult to know how to set up a clinical trial that can test these claims in the first place, unless we are lucky enough to find fairly large groups of patients who all exhibit the same combinations of symptoms and situations.

We will address these worries about the idiosyncrasies of individual patients later in the chapter. As things stand, there have been plenty of placebo-controlled trials of homeopathic treatments, which have proceeded on the usual assumption that it is reasonable to evaluate the efficacy of a particular kind of homeopathic pill for a particular kind of condition.[23] The British Homeopathic Association cautions that "such trials are capable of quantifying efficacy of the homeopathic 'drug' under investigation, but they may yield results that are of questionable relevance to the practice of homeopathy in the 'real world.'"[24]

What have these trials taught us? On the face of things, the picture is mixed. Some investigators have reported a positive benefit of homeopathic remedies beyond that which can be explained by placebo, others have suggested that homeopathic remedies are of benefit equivalent to standard treatment, still others have come to the verdict that homeopathic remedies are no better than placebo. It is important to remember when presented with this mixed picture of the research landscape that not all studies are of the same quality. A very recent survey of all such trials and reviews, conducted on behalf of the Australian government in 2013, has come to the stark conclusion that

> there is a paucity of good-quality studies of sufficient size that examine the effectiveness of homeopathy as a treatment for any clinical condition in humans. The available evidence is not compelling and fails to demonstrate that homeopathy is an effective treatment for any of the reported clinical conditions in humans.[25]

The survey finds, consistently, that the largest and most careful studies show no benefit to homeopathic remedies beyond placebo, while the studies that do suggest a benefit are typically too small to have genuine significance or are flawed in some other way.

Does a verdict of this kind mean that homeopathy has no place in modern medicine? That conclusion would be too quick. Some of the most interesting questions raised about homeopathy in recent years have concerned the degree to which these sorts of methodologically rigorous studies should hold sway over clinical decision-making. Practitioners of homeopathy seem to have the fathers of evidence-based medicine on their side here. Consider this representative statement, from leading proponents of the shift toward basing medical decisions on rigorous evidence, about what they are *not* advocating:

> Evidence-based medicine is not "cookbook" medicine. Because it requires a bottom-up approach that integrates the best external evidence with individual clinical expertise and patients' choice, it cannot result in slavish, cookbook approaches to individual patient care. External clinical evidence can inform, but can never replace, individual clinical expertise, and it is this expertise that decides whether the external evidence applies to the individual patient at all and, if so, how it should be integrated into a clinical decision.[26]

The most interesting site of debate is not so much over the level of evidence accorded to homeopathy; instead, it is over the question of how far the care of individual patients by doctors should be dictated by this evidence base. The British Homeopathic Association insists that "in homeopathy, treatment is usually tailored to the individual. A homeopathic prescription is based not only on the symptoms of disease in the patient but also on a host of other factors that are particular to that patient, including lifestyle, emotional health, personality, eating habits and medical history."[27] Homeopathic practitioners, it seems, are merely exercising their judgment in concentrating on the needs of patients as individuals, rather than adopting the "cookbook" approach rejected even by the medical mainstream.

That said, one might still wonder how homeopathy can possibly have a role in a more "holistic" form of medicine, which attends to the needs of the patient understood as an individual person rather than as a generic locus of symptoms. There surely are cases where a full understanding of the peculiar situation of a patient means that the treatment with the strongest evidence base is not the right one to use. Perhaps an elite athlete, at the end of her career, is so desperate to compete in one last Olympics that she would rather take painkillers than have an operation, even though the outcomes associated with the former are potentially crippling. A responsible doctor, convinced of her patient's personal priorities, might decide that in these circumstances painkillers are the right thing to prescribe. The importance of clinical judgment that is attuned to the idiosyncrasies of the individual patient must not be ignored, but how could clinical judgment ever result in the prescription of homeopathic remedies?

True enough, plenty of people report getting better after using such remedies, and the evidence does not deny their claims. People really do feel better after taking placebos. Maybe some doctors feel, on the basis of witnessing many improvements like these, that their personal judgment of what is required for the unique person in the consulting room is sometimes more powerful than the abstract recommendations issuing from controlled trials for how to treat typical patients. How, though, could one ever be in a position to prescribe a homeopathic remedy if the Australian study mentioned above is right in asserting that the evidence "fails to demonstrate that homeopathy is an effective treatment for any of the reported clinical conditions in humans"?

The answer depends on how we understand the placebo effect itself.[28] The practice of trialing new treatments against placebo can encourage us to write off the placebo effect as insignificant. But the placebo effect is not negligible, nor is it uniform. Taking a sugar pill can make someone feel better. Taking four sugar pills can make that person feel better still. Broadly speaking, the intensity of the placebo intervention correlates with its efficacy: if someone receives an injection of a saline solution, the invasive nature of the process gives it an effect stronger than swallowing pills; large pills seem to have a greater effect than small pills; and capsules are more efficacious than tablets.[29] The very processes of consultation and conversation with a doctor also seem to be efficacious in relieving ill health, and the more elaborate and formalized the processes, the stronger the beneficial outcomes.

We might describe all of this as "mere placebo," but we might also say—without distortion—that the processes of seeking consultation, engaging in attentive discussion with a concerned professional, and ultimately consuming a preparation carefully designed for one's unique set of symptoms can yield significant improvements to health. The benefits that arise from consulting at length with a homeopathic practitioner may even exceed those arising from a quick five minutes with a practitioner of mainstream medicine, especially if the patient in question is suspicious of that latter tradition. We need to remember the "nocebo" effect—placebo's malevolent twin—whereby people who expect poor outcomes from medical encounters end up experiencing precisely those outcomes because of those expectations. Under these circumstances, even a sugar pill can do harm.[30]

It is, of course, entirely irresponsible for homeopathic practitioners to discourage patients with serious conditions from taking medicines that are known to have significant positive effects on those conditions, or to discourage patients from seeking diagnoses from mainstream doctors. Either of these approaches can result in significant harms to patients, even death. But consider a condition like mild or moderate depression. It has frequently been suggested by high-quality clinical trials that much of the positive effect of mainstream medicines on this condition is entirely accounted for by placebo.[31] (The same cannot be said for more extreme forms of depression.) Suppose, then, that a patient has symptoms indicative of moderate depression. Suppose, also, that the patient has had traumatizing experiences with mainstream doctors in the past—experiences that have led to a deep suspicion of mainstream medicine and given him significant hopes for alternative therapies. An antidepressant may not have its usual beneficial effect on such a patient, owing to a strong nocebo effect canceling out the usual significant placebo component of its efficacy. The antidepressant may also have adverse side effects. A homeopathic remedy, presented as such, may instead have a significant positive placebo effect, with no side effects and no damaging nocebo effect.

What is more, the practitioner who takes time to discover a patient's fears regarding mainstream medicine has performed a genuine therapeutic service for the patient. Our understanding of placebo and nocebo justify the verdict that when considering specific patients, with their own idiosyncratic sets of expectations and conditions, some may benefit more from homeopathic remedies than from mainstream treatments. The damning assessment of the aforementioned Australian report leaves the door open for the responsible and judicious use of homeopathic remedies in idiosyncratic circumstances.

A logical follow-up to our discussion of the rights and wrongs of homeopathic therapies is a discussion of the ethics of prescribing placebos. In particular, we might wonder if prescribing placebos is deceptive. The answers to this multifaceted question are not clear, and their detailed assessment must be left for another day. For now, it is enough to note recent research indicating that the placebo effect persists even when people are aware they are taking placebos.[32] In other words, placebos can be presented as such and still be efficacious. Moreover, people typically do not want to know everything about the processes that doctors use to make them better: Why, then, should a patient not rest content when the doctor gives her a placebo, saying quite truthfully that "here is a pill that has worked well for people with your sort of condition, even though we are not quite sure of the mechanism by which it works"? This is exactly how many doctors across the world already make use of placebos in their practice.[33]

Questions about the standing of homeopathy, like questions about the standing of economics and intelligent design, are not best answered by reaching for a single criterion of demarcation between science and pseudoscience. We need different tools to assess the credentials of these very different enterprises, but these diverse tools are sufficient to calm the fear that, when it comes to science, anything goes.

Further Reading

For a brief introduction to economics and its diversity, see:

Ha-Joon Chang, *Economics: The User's Guide* (London: Penguin, 2014).

On the status of economics as a science, see:

Daniel Hausman, *The Inexact and Separate Science of Economics* (Cambridge: Cambridge University Press, 1992).

Alexander Rosenberg, *Economics: Mathematical Politics or Science of Diminishing Returns?* (Chicago: University of Chicago Press, 1994).

For a balanced collection of views on intelligent design, see:

Michael Ruse and William Dembski, eds., *Debating Design: From Darwin to DNA* (Cambridge: Cambridge University Press, 2004).

As far as I am aware, no good philosophical book on the subject of homeopathy currently exists. There is, however, a fine overview of issues relating to evidence-based medicine that includes a detailed discussion of the nature of placebo:

Jeremy Howick, *The Philosophy of Evidence-Based Medicine* (London: Wiley/BMJ Books, 2011).

Chapter Three

The "Paradigm" Paradigm

Popper Versus Kuhn

Students who approach the philosophy of science for the first time usually begin by meeting, and then dismembering, the views of Karl Popper. We did the same in Chapter 1. They then move on to acquaint themselves with the philosophical image of science put forward by Thomas Kuhn. The two thinkers are often cast as great rivals who offer markedly contrasting accounts of scientific achievements and the nature of change in the sciences. Popper takes the role of the champion of scientific rationalism, and of scientific progress. We have already seen how keenly scientists, glad to find a philosopher who massages their collective scientific ego so effectively, have embraced Popper's views.

Kuhn, on the other hand, deals in ideas that seem far more threatening to cherished notions of the advancement of science. It is commonplace to read that Kuhn denies that changes in scientific thinking are rational, and it is even more common to read that Kuhn denies that science makes progress. He is sometimes accused of reducing changes in accepted scientific wisdom to an irrational form of herding behavior, or "mob psychology." It is maybe not surprising, then, that he has been treated with suspicion from many within science.

These efforts to set Popper and Kuhn against each other rely on significant distortions of their writings. It is worth being clear about this at the outset: Kuhn does believe that science makes progress; Kuhn does believe that changes in scientific theory are rational. Indeed, a proper understanding of Kuhn's work shows that his views are far less exotic, and far more persuasive, than a superficial reading suggests. Meanwhile, Popper, who (as we saw in Chapter 1) ultimately grounds the foundations of scientific thought in collective convention, is perhaps more vulnerable to accusations of irrationality and mob psychology.

Thomas Kuhn (1922–1996)

Thomas Kuhn entered Harvard University in 1940 as an undergraduate specializing in physics. In 1945 he began doctoral research—still in physics, still at Harvard—but his interests extended well beyond his thesis topics of quantum mechanics and magnetism. At the time he started his PhD, he simultaneously undertook work in philosophy. He served as editor of the Harvard newspaper *The Crimson,* and he was president of the literary Signet Society.[1] From the late 1940s up until 1956, Kuhn taught a course at Harvard that was intended to familiarize undergraduates in the humanities with work in the sciences. This was when he first became engaged with the history of science, because his teaching method focused on historical case studies going back to Aristotle. In 1956 Kuhn moved to a position in the philosophy department at Berkeley, California, albeit a position in the history of science rather than the philosophy of science.

It was here that Kuhn began to grapple with philosophical work by the likes of Ludwig Wittgenstein and Paul Feyerabend.

Kuhn's best-known work by far is *The Structure of Scientific Revolutions* (henceforth *Structure*), a book that is short, engaging, and important. It was first written in 1962, for incorporation in a series called The International Encyclopedia of Unified Science. This venue for *Structure*'s first publication is ironic, for Kuhn's views are usually thought antithetical to the notion that science as a whole constitutes a unified edifice. Kuhn left Berkeley for Princeton in 1964, and then moved again to MIT in 1983. Much of his later work was devoted to clarifying, modifying, and applying the ideas initially presented in *Structure*: at the time of his death in 1996, for example, he was working on a book exploring an evolutionary conception of the growth of scientific knowledge, an idea that he had first defended in *Structure* itself.

The Structure of Scientific Revolutions

Structure's central thesis is that scientific change is cyclical. Long periods of "normal science," when communities of investigators are more or less united in a vision of what good research looks like, are punctuated by occasional violent conceptual "revolutions." Kuhn contends that examples of these revolutions include acceptance of the idea, following work by Nicolaus Copernicus in the sixteenth century, that the Sun (rather than the Earth) was at the center of the universe and acceptance of Einstein's introduction of the relativistic view of space and time at the beginning of the twentieth century.

Revolutions, says Kuhn, are preceded by a buildup of "anomalies"—problematic phenomena that the anointed scientific approach is unable to account for, no matter how hard scientists try to shoehorn them into accepted explanatory frameworks. After a revolution, scientists embrace a new approach that is able to account for the anomalies that provoked the crisis. Kuhn suggests that scientific communities may need to change their membership for this to occur: sometimes the only way a new approach can gain hold is when the old guard retire from their posts, or when they die.[2] A new period of "normal science" begins, until eventually there is another accumulation of anomalies, another crisis, another revolution. That, in rough terms, is Kuhn's image of science. But what does it involve in detail?

In what Kuhn calls the "pre-paradigm" phase, scientific disciplines are characterized by considerable disunity among their practitioners, often coupled with explicit theoretical debate about the proper foundations of their enterprise. There is little agreement about the requirements of proper scientific training, and little agreement about what sort of thing counts as a significant achievement on the part of earlier thinkers. My own discipline of philosophy is, and most likely always will be, in a state rather like this: there is plenty of valuable activity in the world's philosophy departments, but academic philosophers are not sure about whether their discipline should be directed at examining the history of great philosophical works, exposing the meanings of various problematic concepts, unearthing fundamental facts about the nature of the universe, offering a critical synthesis of the significance of scientific research, or something else altogether. There is also profound disagreement about what counts as good philosophical work. For some, Wittgenstein is a pernicious anti-philosopher who has wrought great damage on the discipline; for others, Wittgenstein is the only thinker to have diagnosed the mistakes of the Western philosophical tradition. Some think of Jacques Derrida's work as groundbreaking, others consider him a charlatan.

When fields of scientific knowledge first got going, Kuhn says, they all had this pre-paradigmatic character, symptomatic of philosophy today. This may be no coincidence, for many—perhaps all—of today's scientific disciplines started out life as speculative branches of philosophy itself. Eventually, says Kuhn, fields of inquiry settle into phases of what he calls "normal science," guided by a paradigm.

This word, *paradigm,* has been used so widely in the management-speak of recent years that we must be careful not to let it wash over us. Instead we must attend to precisely what Kuhn means by it. In the important "Postscript" that he wrote seven years after the first publication of *Structure,* Kuhn acknowledged that he had perhaps used the word in as many as twenty-two different senses.[3] I follow Kuhn himself (and also my former colleague Peter Lipton) in thinking that it is particularly important to think of a paradigm in the specific sense of an *exemplar*—that is, an agreed-upon instance of important scientific achievement.[4]

A paradigm, understood as an exemplar, is not a style of thinking, a worldview, or a form of training. An exemplar is instead a particular example of a solution to a scientific problem. It is something that everyone, or more or less everyone, in a scientific community acknowledges as a piece of work to be admired and emulated. For example, Gregor Mendel's work on inheritance in peas was eventually accorded that status by twentieth-century geneticists. Isaac Newton's work in his 1687 book *Principia* was thought of as an exemplar for centuries. And it seems likely that Charles Darwin structured the *Origin of Species* according to Victorian recommendations for how to formulate and defend scientific hypotheses. Those recommendations, in turn, were based on the efforts of Victorian men of science to pinpoint exactly what had made Newton's work so good.[5]

Kuhn's notion of "normal science" is meant to bring out the idea that this type of science is business as usual: scientists within a given discipline know what sort of work they are supposed to be doing because they agree on which past achievements are exemplary. I do not mean to suggest that all scientists in a community work in precisely the same ways: indeed, this is one of Kuhn's key messages when he tells us that science is guided by exemplars rather than by rules.

It is easiest to see the difference between the notion that science is guided by exemplars and the notion that science is guided by rules if we begin by focusing on activities that are quite distinct from science. A group of expert chefs might agree that Ferran Adrià's work in the 2000s at his restaurant El Bulli in Catalonia is an exemplar for elite cooking, while disagreeing about exactly what made his cuisine so good. Hence they might unite in the notion that Adrià's work should be emulated, while diverging considerably in what they think it means to work "just like him." The cooking styles of these disciples will not be uniform. Contrast this with a rule-based approach, which aims to codify in a far more explicit way what is involved in good cooking. Many amateur cooks in Britain try slavishly to reproduce the recipes of Delia Smith by following her every instruction in detail, even down to using the same cookware. Kuhn's point is that while scientists might be united in their admiration of Newton's achievements, this leaves open the question of exactly how a given investigator will understand what it means to work in the manner of Newton's *Principia*. Scientists are guided by exemplars, but they are not shackled by a detailed recipe book telling them how to investigate the world.

This brings us to a second important point about Kuhn's notion of normal science. The first complete sequencing of the human genome—or rather, the draft sequencing of a supposedly representative genome for our species—was a monumental achievement when it was first announced in 2001.[6] Since that time, we have been treated to more detailed data regarding how the human genome varies, and we have also been given whole genome sequences for many other species, including the genome of the dog, the genome of rice, and the genome of the pigeon.[7] For people with the right equipment and training, genome sequencing is no longer a challenge. It would be tempting, then, to think of the initial human genome project as an exemplar and these other projects as instances of normal genomic science. This could give the misleading implication that, for Kuhn, normal science is just "more of the same"—the mechanical application of methods that have been shown to work by earlier scientists of greater stature.

But Kuhn does not mean to imply that normal science— the work most scientists do, most of the time—is uncreative, or algorithmic, or boring, or trivial. Kuhn's view is that scientific creativity often consists in understanding how a new problem posed to us by nature can be seen as similar to a different problem that we already know how to solve. Galileo began by discovering what happens when a ball is rolled down a slope. When it travels back up another slope it returns to very nearly the same height as that from which it was released, regardless of how steep that second slope might be. He then learned to see the swinging motion of a pendulum as similar to the return of a rolling ball to its release height. A real pendulum has a large weight at its bottom end, but the rod or string that the large weight is attached to also swings, and it, too, has mass. The Dutch natural philosopher Christiaan Huygens later saw that it would be possible to understand the detailed motion of the whole pendulum as if it were composed of a series of connected pendulums, arranged along the line of the string, or rod. In other words, he learned to see a single real pendulum as a collection of simpler Galilean pendulums. Huygens treated Galileo as an exemplar, and Kuhn thinks of Huygens's work as a piece of normal science because of that.[8] But Kuhn also regards Huygens's work as creative, insightful, and important. Normal science is the artful adaptation of that which we already understand to that which we do not.

After a time, normal science may enter what Kuhn calls a "crisis" phase. In a crisis, problematic phenomena begin to accumulate, which no amount of creative work in the style of the agreed exemplars seems able to account for. Science enters a phase of self-doubt. Since scientists are no longer confident that recognizable styles of work will suffice to account for these troubling phenomena, they stop working in carefree emulation of their exemplars and begin to speculate about what proper scientific method should be like, and whether their exemplars have been correctly interpreted. In other words, they spend less time doing science and more time doing philosophy. Eventually, a new theory emerges, often fashioned by younger scholars who are not so enamored of the established exemplars. If this new theory can account for the anomalies left unexplained by previous theorizing, then eventually those old exemplars are cast off and new ones are anointed. A new phase of normal science begins. A scientific revolution has occurred.

What sort of episode does Kuhn have in mind when he describes the general pattern of a scientific revolution? Isaac Newton thought that space was a kind of substance—an infinitely large container in which events might take place. His contemporary Gottfried Leibniz argued against this conception: on Leibniz's view there are physical things—a table, a chair—and we can say how they are related to each other spatially—the chair is one yard to the left of the table—but there is no need to think of space itself as a containing substance.

The Newtonian image of space as a substance seemed to receive a significant boost as later nineteenth-century physicists increasingly accepted the idea that light consists of waves. Sound waves travel through vibrations in air molecules: that is why sound cannot be transmitted through a vacuum. Waves in the sea travel via the up-and-down motion of water molecules. What material medium vibrates when light waves move from one place to another? Not air, for light can travel through a vacuum. It seemed to these physicists that light must travel through oscillations in the substance of space itself, a material without mass that they called the *luminiferous aether*.[9]

The problem was that numerous experiments of increasing ingenuity, designed to detect the luminiferous aether as the Earth moved through it, all failed, or at least they did not yield a decisive verdict in the aether's favor.[10] The aether had become an anomaly; it was something that dominant theories seemed committed to, and yet it could not be detected in any way. And in 1905, with the publication of Einstein's special theory of relativity, physicists converted very quickly to a view of light and space that did not require the aether, and which, more generally, did not require a Newtonian conception of space as an infinitely large container in which physical events are situated. Einstein had effected what Kuhn would call a scientific revolution.

Incommensurability

Kuhn's language of scientific revolutions evokes images in the reader of religious conversion. Perhaps for that reason, Kuhn is often characterized as someone who thinks that the seismic changes in theory that accompany scientific revolutions are irrational: the scientist, it seems, must take a leap of faith, from the old worldview to the new. That impression is further encouraged by one of Kuhn's most notorious assertions—namely, that theories within different paradigms are *incommensurable.* Kuhn himself flatly denies that scientific theory change is irrational, but we cannot understand why he does so until we see what Kuhn means by this notion of incommensurability.

What are the marks of a good piece of scientific theorizing? And how are we to decide when one theory is better than another? As we have seen, Kuhn takes it that normal science is guided by shared exemplars. In endorsing a given piece of scientific work as exemplary, a community of scientists holds up *that* publication—Newton's *Principia,* Darwin's *Origin of Species,* Mendel's work on peas—as setting the standard for quality. If Kuhn is right that exemplars set standards in this way, and if he is right that exemplars change during scientific revolutions, it immediately follows that the very question of what counts as a good piece of science will change after a revolution. This is what Kuhn means when he says that changes in theory across revolution are incommensurable: they have no common measure by which to assess their merits, because standards are informed by exemplars, and exemplars are not constant.

Kuhn thinks that exemplars determine scientific standards in a variety of ways. He is quite emphatic that some very general criteria for assessment persist across scientific revolutions: scientists across all times prefer theories that predict phenomena with accuracy, they prefer theories that are simple, they prefer theories that are plausible in the light of what is already part of established scientific knowledge, they prefer theories that are consistent. Even so, let us focus on just one of these general criteria for quality. What do we mean when we say that a theory is simple? Do we mean it is easy to work with? Do we mean it asserts the existence of very few new theoretical entities? Do we mean the relationships the entities stand in can be modeled using equations of an elegant form?

What is more, the virtues that persist across revolutions will rarely all pull in the same direction. Suppose we must choose between two theories. One is mathematically elegant but seems highly implausible in the light of existing knowledge. Another fits well with what is already known but can be stated only with ugly equations. Which theory should we prefer? Does simplicity trump plausibility, or vice versa? Kuhn's idea is that a scientific community's commitment to following one particular set of exemplars will inform these issues of interpreting the meaning of individual standards and deciding how to balance competing standards against each other. It seems, then, that there is no neutral way to assess, for example, the standing of quantum theory as it was put forward in the early twentieth century. Part of what was at stake was whether its predictive power should override difficulties in understanding what it might mean and how it might be integrated with other areas of physics. Different scientific traditions weigh these factors in different ways.

These are the sorts of themes Kuhn stresses with his talk of incommensurability, but he is careful to limit their significance. When scientists disagree, Kuhn claims that logic will not tell them which theory is to be preferred over the other. There is no deductive procedure that determines, for example, how simplicity should be understood and how simplicity should be weighed against plausibility. Kuhn does not conclude from this that scientific theory change is irrational, or akin to a blind leap of faith. Instead, his claim is that when scientists make these decisions, they employ a form of skilled judgment, one that cannot be understood as the mechanical application of a logical algorithm. This sort of skilled judgment can be rational and reasonable, and ultimately it can sway dissenters.

Suppose I measure the height of my two children using two different rulers. I discover that one is 120 centimeters tall, the other is 3 feet and 2 inches. Which is the taller? Evidently the fact that one height is recorded in the metric system, the other in imperial, does not pose too much of a problem for my comparison, for I need only translate them into the same units. Similarly, one might imagine that so long as we can find a way to translate the findings of one paradigm into the language of another, we will be able to compare them bit-by-bit. We will have no problem in judging Einstein's system as superior to that of Newton, for we can offer an interpretation of Newton's work in Einstein's language.

Especially in Kuhn's later work, he regularly expresses the notion of incommensurability in terms of the limits of translation.[11] He illustrates these problems using the example of the French adjective *doux*. It is hard to make a case for our ability to translate that term *perfectly* into English.[12] While a French speaker calls a pillow *doux*, an English speaker would say it is soft; while he calls butter *doux*, an English speaker would say it is unsalted; while he calls wine *doux*, an English speaker would say it is sweet; while he calls the actions of a child *doux*, an English speaker would say they are gentle. What is more, to the French ear the term *doux* is not *ambiguous*: it is not like the English term *bank*, which has two entirely distinct meanings (namely, the place where you deposit your money, and the place by the side of a river). Instead, *doux* has a single meaning in French, one that is far broader then the meaning of any corresponding English term.

We should agree, then, that a term like *doux* cannot be translated perfectly into English, for no single word in English will bring with it the same broad range of resonances conveyed by the French term. The meanings of key scientific terms like *mass* or *gene* also differ as we move from the theories of Newton to the theories of Einstein, or as we move from Mendel's advocates at the beginning of the twentieth century (who knew nothing of the internal nature of chromosomes), through the work of Watson, Crick, and others on the double-helical structure of DNA, and on to the molecular biology of the present day. Kuhn's thought is that just as we cannot convey the full content of French judgments about unsalted butter using English, so we cannot convey the full content of Newton's outlook using the language of Einstein.

Once again, at the same time as stressing that the impossibility of perfect translation contributes to the incommensurability of distinct paradigms, Kuhn is also careful to contain the significance of this point. Even if French cannot be translated perfectly into English, it is possible for French and English people to communicate with each other adequately, and it is possible to formulate serviceable English translations of French texts. What is more, the failure of perfect translation does not destroy the ability of French and English speakers to disagree with each other, and to settle their disagreements to the satisfaction of both parties. If I am convinced that the waiter is going to bring us salted butter, but my French friend Philippe thinks the butter will be *doux,* then we can decide who is right by tasting some when it arrives. Likewise, Kuhn says that in spite of the fact that two scientists operating within different paradigms cannot translate their work perfectly into the language of the other, this does not mean they cannot understand each other, and it does not mean that they cannot devise experimental procedures that will determine, to the satisfaction of all parties, which paradigm is the better.[13]

Different Worlds

Kuhn does not think that scientists are trapped in bubbles inflated by their own theorizing, which prevent them from understanding, talking with, or persuading the occupants of alternative theoretical bubbles. For the most part his detailed views are altogether more sober. That said, things get more exotic in *Structure*'s famous tenth chapter, for here Kuhn argues that revolutionary changes from one paradigm to another have the most profound effects imaginable.

Kuhn's own early experiences of delving into ancient works of science led him to the view that the universe itself is transformed for investigators working in different paradigms. When Kuhn was preparing his first lecture course in the history of science, he read Aristotle's *Physics* (a work written in the fourth century BC) in a naive effort to find out "how much mechanics Aristotle had known, how much he had left for people like Galileo and Newton to discover." At first, the entirely unsurprising conclusion that Kuhn came to was that, in spite of his formidable reputation, Aristotle had known nothing of modern science. Worse, Aristotle's work was incomprehensible and incompetent. But after a little time mulling over Aristotle's claims, Kuhn experienced a revelatory transformation in his vision:[14]

> I was sitting at my desk with the text of Aristotle's *Physics* open in front of me and with a four-colored pencil in my hand. Looking up, I gazed abstractedly out of the window of my room—the visual image is one I still retain. Suddenly the fragments in my head sorted themselves out in a new way, and fell into place together. My jaw dropped, for all at once Aristotle seemed a very good physicist indeed, but of a sort I'd never dreamed possible.

Much later in *Structure,* Kuhn would generalize from his personal Aristotelian gestalt-shift, telling his readers that "after a revolution, scientists work in a different world."[15]

It is primarily because of remarks such as this one that Kuhn has been called a *relativist.* He seems to be telling us not merely that scientific ideas about the world change when one theory replaces another, but that the world itself—the very object science seeks to investigate—changes with that revolution. On this view, competing theories do not offer alternative understandings of the same universe; instead, the nature of the universe depends on the theory used to describe it. Why would Kuhn say such a thing?

It is not always clear whether Kuhn does say anything quite so radical, for the language he uses slips between mild and strong claims:

> Do we, however, really need to describe what separates Galileo from Aristotle, or Lavoisier from Priestley, as a transformation of vision? Did these men really *see* different things when *looking at* the same sorts of objects? Is there any legitimate sense in which we can say that they pursued their research in different worlds?[16]

Here Kuhn asks whether two scientists in the grip of different theories literally see things differently, or whether instead they see things in precisely the same ways, while coming to different conclusions about the significance of what they see. Kuhn thinks we should embrace the first option: he thinks theoretical commitments make a difference to how we see things. His argument draws in large part on work in the psychology of vision. If you put on special goggles that invert the image of the world arriving at your retina, initially everything will appear upside down. You will be disoriented and clumsy. But after a while you learn to compensate for the odd effects of the goggles, and things will look just the same as they did before you put the goggles on. Once habituated in this way, you will find that it is only when you remove the goggles that things once again seem the wrong way around.

Kuhn is on safe ground, then, in thinking that our visual experience is plastic; that is, how we see things can be altered over the course of our lives. More specifically, it can be transformed by our beliefs: if a few playing cards in a standard deck are doctored so that, for example, the queen of hearts is black, or the four of spades is red, then, so long as people are exposed to them reasonably briefly, they will not notice anything untoward, and will instead identify these anomalous cards as a normal red queen of hearts or a normal black four of spades. Our expectations for how things are—in this case, our familiarity with a standard deck of playing cards—make a difference to how things appear to us.

Technical training can also affect how things look: as the philosopher Ian Hacking has stressed, whereas the layperson looks at an X-ray image and sees only blobs, some of which may be suggestive of bone, the experienced doctor looks at the same image and a diagnosis leaps from the picture. She sees a tumor where we see nothing, or just a blur.[17] Kuhn's view, then, is that training and theoretical convictions make a difference not just to the conclusions scientists draw from their microscope slides, or from their telescopes, but to how they see the world they investigate with these instruments. Even so, there is a significant leap from Kuhn's mild notion that two scientists "see different things," in the sense that things look different to them, to the far stronger notion that they literally pursue their research "in different worlds."

Occasionally we get the impression that Kuhn's assertion that different scientists work "in different worlds" is merely a colorful way of making vivid his more basic, and far less contentious, conviction that the world starts to look different when your theoretical commitments have changed. But Kuhn's claim about the world changing after a scientific revolution is more than just a *façon de parler*. To understand why, we need to examine the appeal that the views of the eighteenth-century German philosopher Immanuel Kant held for Kuhn.

Kuhn's Kantianism

By and large, people agree on what colors things are. Most of us would say that ripe tomatoes are red and that grass is green. Sometimes we make mistakes about color—perhaps we look too quickly, perhaps we are viewing things under peculiar forms of illumination—but still we can correct ourselves by looking more carefully, or by taking objects into a source of natural light. In spite of all this, many scientists and philosophers (although certainly not all) would argue that colors do not exist in objects themselves.[18] Instead, they hold that colors are artifacts of human visual perception. Colors are something that objects *appear* to have, but this appearance is merely a consequence of how the human visual system processes information arriving at the eyes. Colors, on this view, are not genuine properties of material things. Nonetheless, because humans largely share the same perceptual systems, we have fairly robust standards for what count as the "true" colors of objects.

The nature of color, on this view, is not something that exists independently of experience. As a *very* rough simplification, we can say that Kant had similar thoughts about space and time. They, too, said Kant, are not features of the universe that exist independently of human experience. Kant thought that this radical proposal helped to explain some puzzling features of geometry. Up until the end of the nineteenth century, Euclidean geometry was widely thought to give an accurate description of the nature of space. But Euclidean geometry also seems to be an activity that one can do entirely from the armchair: one does not need to set up experiments to show that the angles of a triangle add up to 180 degrees. How is it possible that a science can, at one and the same time, tell us about the nature of space and yet demand no significant interaction with the world? Why don't we need to do experiments to determine the nature of space? Kant's idea, defended in his 1781 work *The Critique of Pure Reason,* was that this puzzle could be resolved if we thought of the properties of space as, in some sense, arising not from nature in itself but from how humans experience things.

Kuhn embraces a form of Kantianism. For Kuhn, the world itself does not exist independently of the way we experience it, and, as we have already seen, he also believes that the way we experience the world is affected by our scientific theories:

> As a result of discovering oxygen, Lavoisier saw nature differently. And in the absence of some recourse to that hypothetical fixed nature that he "saw differently," the principle of economy will urge us to say that after discovering oxygen Lavoisier worked in a different world.[19]

Just as many philosophers have been tempted to deny that the world contains *real* colors, understood entirely independently of the way human perceivers tend to see them, so Kuhn sees no reason to posit a *real* world that is independent of the way human scientists tend to see it. Of course, given that most humans do tend to see colors in similar ways, it makes sense to say that someone has made a mistake if he tells us that grass is purple. But these standards of correctness are relative to human vision in general. Different species have different visual systems, giving rise to different capacities for visual discrimination and classification of surfaces. Most humans have three types of cone cells in their eyes (although some color-blind people have only two), whereas goldfish have four, and pigeons have five.[20] It is hard to know, then, what we might mean by talking of the *true* color of a flower, if that is to be understood independently of the species of organism that happens to be looking at it.

We have seen that Kuhn stresses that scientists see the world differently before and after revolutions. He thinks of this as akin to a shift in their perceptual systems. If we are talking about scientists who share a paradigm, Kuhn is happy to say that some have gotten things right while others have got it wrong. But he denies that there is a way things are, independent of all scientific theorizing. Just as standards for correctness in color attribution are species-relative, Kuhn thinks that standards for the correctness of claims about the world are paradigm-relative. That is why Kuhn thinks that the worlds in which scientists work change with paradigm shifts.

Evolutionary Progress

Kuhn's Kantianism also explains his views about scientific progress. One might think of progress in science as the provision of an increasingly detailed picture of how the universe is. But Kuhn denies that there is a way the universe is, understood independently of any group of scientists' views about how things are. In that sense, the universe is not a stable object of investigation, which science might eventually capture. Instead, in Kuhn's view, the universe is a moving target: as our paradigms change, the universe changes, too.

Kuhn cannot claim that scientific progress consists in gradual convergence over time on stable facts about our universe, for he denies there are stable facts about the universe. How, then, can Kuhn make sense of progress at all? In *Structure's* final chapter, entitled "Progress Through Revolutions," Kuhn invokes Darwin to illustrate his views. Kuhn hopes that an analogy with Darwinian evolution will help him explain to readers what progress might mean, if it is not progress toward some stable form of truth. Kuhn contends that Darwin, too, thought that evolution was progressive, and that Darwin, too, thought that evolutionary processes do not begin with some stable goal, specified in advance.[21]

Suppose we ask "How should a species ultimately end up, if it evolves by natural selection in a grassland environment?" There is simply no good answer if we pose our question in such a bald way. Even if we think that natural selection leads to progress via slight improvements, the question of what an improvement might look like in such an environment depends on whether we are talking about a large grazing mammal, an insect parasite, or a bird of prey. Moreover, the grassland environment itself is not fixed: species change their environments as they eat grass, as they produce dung, as they decompose, as they breathe.[22] Our question is a bad one, in part because we cannot say what counts as a forward move in the evolutionary game unless we specify what sort of a species we are talking about, in part because the environment of any species is a moving target.

Kuhn's idea, based largely on his Kantianism, is that when we ask what science is meant to conform to, we find that the universe it seeks to describe is also a moving target; and when we ask what counts as an improvement to a scientific theory, the answer depends on how that theory construes the world. Even so, says Kuhn, just as it makes sense to think that natural selection favors those organic variants that are slight improvements on what went before, so scientific communities prefer the theories that offer better solutions than their predecessors to the problems they address. Kuhn rejects the notion that science provides an increasingly accurate picture of a world whose structure is independent of what we happen to think about it. Still, science makes progress. Finally we can understand why Kuhn looked back on his work and described it "as a sort of post-Darwinian Kantianism."[23]

Evaluating Kuhn

This chapter has aimed to encourage an understanding of, and sympathy for, Kuhn's image of the processes of science. How well do Kuhn's views hold up?

For Kuhn, normal science and revolutionary science are very different in kind. Normal science consists of what he calls "puzzle solving"—that is, taking on problems, confident that the creative adaptation of respected exemplars will eventually yield solutions. After a revolution, the old exemplars are rejected, and new ones anointed. Kuhn says that when—and only when—revolutions happen, worlds change. In spite of the considerable ingenuity and importance that accompany innovations in normal science, discoveries of this more modest sort leave the world intact.

If revolutionary science and normal science are qualitatively distinct in these ways, it had better be the case that we can tell if we are dealing merely with an exceptionally insightful piece of normal science, or if instead we are in the presence of a revolutionary *bouleversement*. While that distinction may seem intuitive enough when we are talking about theories of the cosmos itself—revolutions occur when the Earth is deposed from the center of the solar system, or when Newton is deposed in favor of Einstein—it is far less clear how we are supposed to apply Kuhn's scheme once we look away from physics and toward other sciences such as biology.

By any standard reckoning, Darwin's *Origin of Species* is an exemplary scientific work.[24] It is unusual in being read regularly by practicing biologists today, in spite of the fact that it is over 150 years old. When biologists squabble over contentious scientific issues, they often try to recruit Darwin to their team. But although Darwin's work is important, it is not clear that its publication amounted to a revolution in Kuhn's sense. And yet, if Darwin's work does not count as revolutionary, we must question whether Kuhn's distinction between normal science and revolutionary science can be applied in biology at all.

Soon after Darwin's book was published in 1859, natural historians quickly converted to the "transformist" view defended in that work. In other words, they were quickly persuaded that the species we see around the world are descended from a small number of common ancestors, which had undergone a series of gradual change over vast stretches of time. It would be tempting, then, to think Darwin's work must have been revolutionary in character, on the grounds that it effected a wholesale shift in how the organic world was understood. But Darwin was certainly not the first to suggest that distinct species might be related genealogically, and he was not even the first to provide evidence for this. The same idea had been tabled by French naturalists such as the Comte de Buffon and Geoffroy St. Hilaire earlier in the eighteenth and nineteenth centuries.[25] Transformism was an idea in common circulation in scientific circles, and the anonymous publication in Britain in 1844 of *Vestiges of the Natural History of*

Creation—fifteen years before the *Origin* was published—made it an idea widely discussed among the general public.[26]

Darwin's work had a swift impact on the scientific community, but it did so by marshaling a mass of evidence, of diverse sorts, in favor of transformism, and by laying out a persuasive case in its support. Darwin did an enormous amount to make transformism respectable and compelling to the scientific elite. While this outcome allows us to say that Darwin brought about significant changes in received scientific thinking, it does not mean that Darwin's work was revolutionary in the Kuhnian sense. Transformism was not remotely alien to natural historians who read the *Origin* when it first appeared.

While transformism was not a new idea, natural selection was. Darwin put it forward as a novel explanation for the exquisite adaptations we see in plants and animals. This part of Darwin's theory was distinct from the broader transformist notion that plants and animals are modified descendants of ancestors held in common. Perhaps it is in the formulation of this hypothesis—namely, that species become adapted to their environments through a process of competitive struggle—that the *Origin* earns the right to be considered a revolutionary work in Kuhn's sense?

There are a number of problems with this interpretation. First, although natural selection was a new idea, it was formed by the creative fusion of many old ideas that would have been familiar to Darwin's readers. Darwin presented natural selection as analogous to artificial selection, a phenomenon that all of his contemporaries would have known about via the conspicuous successes of animal breeders like Robert Bakewell in improving cattle and sheep.

Darwin argued that anything the breeder can do on the farm, nature can do better in the wild. He claimed that this "selection" was achieved as a consequence of wild populations expanding in a way that outstrips available food resources, with the result that only the very best adapted would survive. That idea, too, would have been familiar to those who, like Darwin, had read Thomas Malthus's *Essay on the Principle of Population,* published in 1798. It is hard to know, then, whether we should understand Darwin as combining preexisting elements of respected work in the innovative manner characteristic of normal science, or whether instead we should understand his insight as paradigm-busting. What is more, Darwin was not able to persuade many of his contemporaries that natural selection was an important agent of adaptive change.[27]

It is not difficult to see why Darwin had trouble selling the idea of natural selection. For example, a fairly negative review of the *Origin* by the Scottish engineer Henry Fleeming Jenkin asked why we should be so sure that iterated cycles of variation and selective competition are able to produce increasingly refined adaptations.[28] Why, for example, should we think that Darwin's principle of natural selection can explain, as Darwin assures us it can, increasing running speed in wolves? Suppose that beneficial variations arise rarely. Perhaps a few members of a population of wolves are born who can run a little faster than the others. They have more babies as a result. But because these beneficial variations are rare, the chances are that when one of these faster wolves finds a mate, that mate will run at an average speed. When this couple has a baby, its running speed will probably end up closer to the population average than that of its single speedy parent. This baby, too, is likely to mate with average runners. Over time, says Jenkin, the benefit that initially accrues from faster running will be washed away owing to these repeated cycles of mating with more average specimens.

Darwin thought he had an answer to Jenkin's challenge, but it is very different from the response we rely on today. Darwin thought that slight, beneficial variations were really rather common. He thought that faster-running wolves would regularly appear in the population. He also thought that the struggle for existence was so exceptionally intense that the more average wolves would perish before they could mate. Finally, he thought the tendency to produce faster offspring could itself be inherited, with the result that once selection began to favor fast running, it would amplify the number of wolves who could run even faster still.

Darwin's effort to answer Jenkin's challenge is far from the image we have of selection today.[29] Like Darwin, modern biologists take the view that Jenkin's mistake was to think that beneficial variation would be lost because of repeated cycles of mating. Unlike Darwin, they argue that the nature of genetic inheritance—evidently something Darwin could not have known about—allows beneficial variation to be preserved even when, for example, faster-running wolves mate with others who are more average. It requires refined mathematical apparatus to make this case, and Darwin himself never dealt with complex maths. In the end it was not until the 1920s, with the mathematization of evolutionary theory at the hands of people like the Cambridge statistician and geneticist Ronald A. Fisher, that natural selection began to be widely accepted among biologists as a potent force in evolution.[30]

In retrospect we can look at Darwin's book and say that it made a strong case for natural selection, but in reality natural selection was assured its place in the explanatory toolbox of practicing biologists by much later efforts of Fisher and others. In sum, it is difficult to understand the history of biology in

Kuhnian terms, for it is unclear whether a work like the *Origin* counts as introducing a revolution. Kuhn, remember, was a physicist by training, and his approach struggles to account for the broader diversity of scientific practice. In particular, his framework of grand paradigm shifts seems ill-suited to the explanation of changing theory in biology.

The Plurality of Exemplars

There is a final problem that arises when we try to approach biology in a Kuhnian fashion, and it has broader significance for Kuhn's treatment of exemplars. Kuhn himself seems to suggest that when revolutions occur, the old exemplars are discarded and replaced with new ones. But why should this be the case? After all, an exemplar is a concrete achievement—something to be emulated. Kuhn himself stresses that the mere fact that something is seen as admirable leaves open the question of exactly what makes it admirable, or exactly how it should be emulated. That, in turn, should make us wonder why old exemplars need to be cast aside altogether after a revolution. Might they not instead be continually reinterpreted as they recede further into history?

We have seen that Darwin's detailed account of the workings of natural selection was unlike the framework biologists use today. It was, for example, free of mathematics and it relied heavily on a notion of intense struggle to counter the problems posed by Jenkin. It alleged the inheritance not just of variations but of the capacity to produce variation in a given direction, and of course it made no mention of genes. For the modern biologist, the mathematical treatment of evolutionary processes, set against a background of genetic inheritance, which Fisher put forward in his landmark 1930 work *The Genetical Theory of Natural Selection,* is a primary exemplar of how evolutionary biology should be done. But none of this means that the *Origin* isn't also a primary exemplar, for the *Origin,* too, still offers an inspiring vision of how to construct an evidentially rich account of how species evolve over time—one that has a version of natural selection (albeit not quite the version we have today) at its core.

For Darwin himself, and for his Victorian contemporaries, Newton's *Principia* was also an exemplary work of science—not because Darwin wanted to find biological analogues of Newtonian mass, or Newtonian space, but because Darwin believed that Newton's work showed in general terms how one should go about constructing a persuasive case in favor of a novel hypothesis. Today we no longer think that Newton is right about cosmology—in that sense his work has been displaced—but it does not follow that Newton's work is no longer an exemplar of diligent scientific activity, in the same very general sense that it was exemplary for Darwin. We do not need to cast exemplars aside, even after the eclipse of what might seem, for a while, to be their most important achievements. If exemplars are indeed preserved and reinterpreted across great swaths of scientific history, it becomes harder to talk of wholesale paradigm changes.

There is much to admire in Kuhn's work, especially when it comes to his insistence that exemplars play a role in guiding science in a way that does not reduce scientific activity to the mechanical application of rules. But that does not mean we need to retain what is now Kuhn's most notorious idea. It is time to bring to a close the paradigm of revolutionary paradigm shifts.

Further Reading

The single most important thing to read is, of course, Kuhn's own most influential work. A fiftieth-anniversary edition has recently been published, with a very helpful introductory essay by Ian Hacking:

Thomas Kuhn, *The Structure of Scientific Revolutions, Fiftieth Anniversary Edition* (Chicago: University of Chicago Press, 2012).

Also worth reading is an important collection of later essays by Kuhn:

Thomas Kuhn, *The Road Since Structure: Philosophical Essays 1970–1993* (Chicago: University of Chicago Press, 2000).

For discussion between Kuhn, Popper, Lakatos, and others on matters covered in this chapter, see:

Imre Lakatos and Alan Musgrave, eds., *Criticism and the Growth of Knowledge* (Cambridge: Cambridge University Press, 1970).

Two very helpful books devoted to understanding Kuhn are:

Alexander Bird, *Thomas Kuhn* (London: Acumen, 2001).

Paul Hoyningen-Huene, *Reconstructing Scientific Revolutions: Thomas S. Kuhn's Philosophy of Science* (Chicago: University of Chicago Press, 1993).

For a fascinating, and very recent, study of Kuhn in his historical and institutional context, see:

Joel Isaac, *Working Knowledge: Making the Human Sciences from Parsons to Kuhn* (Cambridge, MA: Harvard University Press, 2012).

Chapter Four

But Is It True?

Nature as It Is

The sciences have been instrumental in many of humanity's most momentous endeavors. They have been used to send people to the moon, to create nuclear weapons, to give women control over their fertility, to construct the personal computer and the Internet. Acute commentators have disagreed over what all these achievements tell us about the image of the world that science yields. Do the sciences give us an accurate representation of how things truly are? Or do they instead offer us something of great importance, but something quite distinct from a true picture of our universe—maybe a set of techniques for computing observations, or a series of detailed stories that, whether true or false, have earned their keep through their remarkable practical value?

Before launching into the debate over science and truth, it is worth making a few preliminary remarks. *Scientific realism* is the label for the philosophical view that science is in the truth business. Scientific realism says that the sciences represent those parts of the world they deal with in an increasingly accurate way as time goes by. Scientific realists are not committed to the greedy idea that the sciences can tell us all there is to know about everything; they can happily acknowledge that there is plenty to learn from the arts and humanities. Moreover, by denying that science gives us a *perfectly* accurate picture of the world, scientific realists are not committed to the manifestly absurd idea that science is finished. The claim that the sciences give us increasingly accurate representations leaves plenty of room for revision and improvement, as more refined images of nature are produced. This means that scientific realism is a position worth discussing: it isn't obviously wrong, but it isn't obviously right, either.

A moment's reflection suggests that scientific realism is not the only sensible and respectful way to respond to the successes of science. Perhaps we should think of scientific theories in the way we think of hammers, or computers: they are remarkably useful, but like hammers and computers they are mere tools. It makes no sense to ask whether a hammer is true, or whether it accurately represents the world, and one might argue that the same goes for science: we should simply ask whether its theories are fit for their purposes. Or perhaps we should think of scientific theories in the same way some prominent Anglicans seem to think about the stories in the Bible: they are stimulating fictions, or maybe they are full-blown falsehoods, but we hold on to them because of the ways in which they help us to navigate the world.[1]

Cutting to the chase, this chapter will argue in favor of scientific realism. But the pathway toward that conclusion is not straightforward, and so a little signposting is in order. There are three jobs we must accomplish if a good case is to be made for scientific realism. First, we need to fend off one of the most potent arguments against scientific realism—namely, the argument from "underdetermination." Roughly speaking, this argument suggests that scientific evidence is never powerful enough to discriminate between wholly different theories about the underlying nature of the universe. The result, says the proponent of underdetermination, is that the body of scientific evidence can never justify the conclusion that our best scientific theories are true, or even close to the truth.

Second, we need to ask whether there is any positive argument in favor of scientific realism. More or less the only argument that has ever been offered to support this view is known as the "No Miracles argument." The basic gist of this argument is that if science were not true—if it made significant mistakes about the constituents of matter, for example—then when we acted on the basis of scientific theory, our plans would consistently go awry. The strongest case to be made in favor of the truth of scientific theory, in other words, draws on the conspicuous successes that the sciences have facilitated.

Third, and finally, we must confront an argument known as the "Pessimistic Induction." This argument draws on the historical record to suggest that theories we now think of as false have nonetheless been responsible for remarkable practical successes. We now think, for example, that Newton's view of space is strictly false in the light of Einstein's relativistic theory. And yet it was Newton's theory that was used, successfully, to send men to the moon. If falsehood, rather than truth, can regularly yield success, then the No Miracles argument is in trouble. And if past theories are consistently written off as false in spite of their practical successes, then the chances are that our most cherished modern theories will turn out to be false, too.

In short, the scientific realist needs to show that considerations of underdetermination are impotent, that the No Miracles argument works, and that the Pessimistic Induction fails. This chapter aims to achieve precisely those tasks, in precisely that order.

Underdetermination

One of the most significant challenges to scientific realism comes from a phenomenon known in the technical jargon of philosophy as *the underdetermination of theory by data.*[2] This intimidating string of words hides a simple idea, for to say that two competing theories are underdetermined by data is simply to say that we do not have enough evidence to decide which is correct.

Situations of underdetermination are not restricted to the sciences. Christopher Clark gives a sense of how hard it can be to find the evidence we need to adjudicate between alternative historical accounts of events in *The Sleepwalkers,* his magnificent account of the origins of World War I:

> Reconstructing the details of the plot to assassinate Archduke Franz Ferdinand in Sarajevo is difficult. The assassins themselves made every effort to cover the tracks that linked them to Belgrade. Many of the surviving participants refused to speak of their involvements; others played down their roles or covered their tracks with obfuscating speculations, producing a chaos of conflicting testimony. The plot itself produced no surviving documentation: virtually all those who took part were habituated to a milieu that was obsessed with secrecy.[3]

As Clark himself is quick to point out, problems such as these are not fatal for the project of historical reconstruction: new evidence can come to light as diaries and letters are discovered, as archives are opened, or simply when known sources are read and compared with greater care. The same is the case in science. Hypotheses that begin as shaky speculations can become better confirmed as new data are produced. Franz Boas, one of the founders of modern anthropology, remarked in a lecture given in 1909 that many of Darwin's claims about human origins, made nearly forty years earlier in his 1871 work *The Descent of Man,* had a rather skimpy basis in data. But time and hard work had given the anthropologist new archaeological and anatomical information that put Darwin's claims on a far firmer footing:[4]

> At the time when Darwin wrote, the evidence bearing upon the various points here quoted was very fragmentary; but the unceasing endeavors to find evidence supporting or invalidating his theories have led to a much better understanding of the problem. We find that much evidence has been accumulated which proves with a fair degree of certainty the close relationship between man and the higher apes of the Old World.

Today we have not only additional fossil finds, but entirely new forms of evidence in the shape of DNA analysis, various dating techniques and so forth, which further magnify our abilities to discriminate between different hypotheses regarding human ancestry. The picture is still far from fully resolved, but it is much clearer now than it was in 1871.

Scientific realism is not troubled by the thought that situations of evidential balance can eventually be resolved one way or another. Indeed, it is hard to see how else science might make progress—how it might produce increasingly accurate representations of the world—unless it is by using experiment, and the painstaking collection of previously unavailable data, to move from reasonable indecision to confidence in the face of choices between competing theories.

Scientific realism can also concede that some questions in the sciences may never receive decisive answers: maybe we will never be able to acquire the sort of data that tell us the color of the bony plates on the back of *Stegosaurus;* maybe the most obscure areas of fundamental physics will forever remain conjectural. For underdetermination to constitute a general threat to scientific realism, it must have the result not that agnosticism is sometimes the right attitude to take to our best scientific theories but that agnosticism is always the right attitude.

Formulating a good argument against scientific realism requires that we state the problem of underdetermination in an especially potent way. We could try to claim, for example, that however much evidence we have gathered in favor of our best scientific theories, the possibility remains that some other theory, which makes radically different claims about the world, might account for the same body of evidence just as well. If this is the state we find ourselves in, then while we may have plenty of practical reasons to continue to use the commonly accepted theory— maybe it is easy to use, or easy to teach, and it helps us to predict events that we care about—we have no reason to think that it is true, or even close to the truth. If this very general challenge from underdetermination works, then it would have the result that agnosticism is the right attitude, even to the very best science has to offer. Scientific realism would be undermined, because our evidence would always be unable to discriminate between fundamentally conflicting images of the universe.

Duhem to Descartes

The historical roots of the underdetermination problem are sometimes traced to the French scientist, and philosopher of science, Pierre Duhem (1861–1916), especially his 1906 work

The Aim and Structure of Physical Theory.[5] Although it is common to see Duhem's views linked to arguments from underdetermination, we should be wary of equating Duhem's reflections on scientific method with the sort of skeptical attitude that usually accompanies strong versions of the underdetermination challenge.

Duhem's primary target was an unwarranted confidence in the power of deduction to tell us whether a hypothesis in physics is mistaken. He pointed out that when our experimental results fail to line up with what our theory tells us to expect from the world, it is possible that the theory itself is at fault, but it is also possible that our apparatus is faulty, or that the assumptions that underpin our use of the apparatus are faulty.

This is precisely the situation we came across back in Chapter 1. There we encountered the surprising observations made at Gran Sasso in 2011, which appeared to indicate that neutrinos had exceeded the speed of light. These results did not immediately cause scientists to abandon the principle, inherited from Einstein, that nothing travels faster than light. Although that principle may have been put into question in the minds of some physicists, the anomalous results from Gran Sasso also raised the questions of whether the experimental equipment had been properly assembled and whether the principles used to calculate the neutrinos' speed were correct. Duhem drew a general moral from similar incidents:

> The physicist can never subject an isolated hypothesis to experimental test, but only a whole group of hypotheses; when the experiment is in disagreement with his predictions, what he learns is that at least one of the hypotheses constituting this group is unacceptable and ought to be modified; but the experiment does not designate which one should be changed.[6]

Duhem is reminding us that an experimental result cannot tell us, all by itself, whether to accept or reject the hypothesis we are trying to test. A puzzling piece of data can tell us that we have made a mistake somewhere, but we need to bring additional considerations to bear when we ask where the mistake has occurred. Duhem drew the conclusion that a good scientist needs more than an experiment if she is to determine which hypotheses to accept and which to reject. She also needs a cultivated sense of good judgment in order to decide whether the blame for peculiar experimental results most likely rests with faulty apparatus, faulty calculations, or a faulty fundamental theory. Duhem did not conclude that a determined scientist might reasonably cling on to any theory she wants to, regardless of how the data turn out, and neither did he conclude that every successful scientific hypothesis has a radically different competitor that is just as well supported by the available evidence. These, however, are the sorts of claims we need to defend if underdetermination is to pose a problem for scientific realism.

Is there a way to make a case for these more threatening forms of underdetermination? Every chemist—and most schoolchildren—will tell you that water is predominantly composed of molecules that contain two hydrogen atoms and a single oxygen atom. This consensus took hard work to achieve: before the 1860s it was thought that the chemical formula of water was HO rather than H_2O, and before the 1780s chemists took water to be an element rather than a compound. Looking back, it would be easy to think that these quaint mistakes should have been obvious even to the chemists who proposed them. This attitude obscures the quite genuine balance of evidence that, when these debates were live, made decisive choice between competing views of water's structure difficult, perhaps even premature.

The evolution in our views about water continues. As microscopy has become increasingly refined, some scientists have recently generated what they take to be images of single H_2O molecules.[7] But even now we are realizing that water is not merely H_2O. As the philosopher and historian of science Hasok Chang points out, a pure heap of H_2O molecules would not be recognizable as water, for the properties we typically associate with water depend on the additional presence of various ions.[8] So is there still an alternative structure we can propose for water that is radically different from H_2O, and which accounts just as well for all the properties we now know water to possess? In the light of all these data, could water molecules be predominantly composed of atoms of silver, or of helium, or of a hitherto unknown element? Or might water not contain molecules at all? Of course, the answer is that *now* we know of no sharply different scientific alternative that accounts for our data just as well as the H_2O hypothesis. This is precisely why chemists agree so overwhelmingly about what the basic structure of water is.

How, then, have underdetermination's champions tried to make a case for agnosticism and against scientific realism? Many have presented underdetermination either as a sort of promissory note, or in a wholly general way. The first strategy points out, in humble acknowledgment of our fallibility, that while we do not have any detailed alternative understanding of the microstructure of water, there could be one that we are not aware of, and it might account for all the known facts about water just as well as the view that water is H_2O. The second strategy instead offers a recipe for cooking up underdetermined hypotheses, such as André Kukla's simple suggestion that "for any theory T, construct the rival T* which asserts that the empirical consequences of T are true but that T itself is false."[9] Applying this recipe to the water case, the alternative to the H_2O

hypothesis is the hypothesis that "everything is as though water were composed of H_2O, except its structure is really entirely different." The opponent of scientific realism concludes that we should not think water is likely to be H_2O, because our evidence cannot decide between these competing alternatives.

These ways of using underdetermination to undermine scientific realism do not trace back to Duhem and his detailed examination of scientific practice. Instead, they have their roots in the work of another, much earlier French philosopher—René Descartes (1596–1650)—and his wholly general meditations on human knowledge.

As I write these words, it seems to me that I am sitting in a train on the way to Leeds. One hypothesis that accounts for the evidence of my senses is that I am indeed sitting in a train on the way to Leeds. A very different hypothesis, which seems to account for the same evidence just as well, is that an all-powerful demon has tampered with my mind, causing me to have experiences just as though I were sitting in a train on the way to Leeds. In reality I am under the demon's power. The evidence of how things seem to me does not discriminate between the two—in other words, the train hypothesis and the demon hypothesis are underdetermined—and so I should withhold judgment on the train-versus-demon issue.

This book—which keeps its focus squarely on the philosophical problems raised by science—is not the right place to address the profound questions raised by this type of generic skeptical challenge. It is, however, the right place to note that champions of underdetermination face a problem if they want to undermine scientific realism. They cannot show that our best scientific theories invariably face detailed, serious, competing views of how the universe is structured. If there were such theories, then scientific agreement would be far less common than it is. The best they can do is suggest, in wholly general terms, that the true nature of the world may conceivably be very different from how our best scientific theories tell us it is. But this is not to raise any particular worry about the status of scientific knowledge: it is merely to point out that an all-powerful demon may be deceiving us about the structure of water, just as she may be deceiving us about our engagement with public transport. The problem of underdetermination is either no problem at all or it is the old-fashioned problem of general skepticism posed by Descartes. In neither case is it a special problem for the standing of scientific knowledge.[10]

No Miracles, Please

If you are trying to navigate a ship through treacherous waters, then unless you make an effort to find out the whereabouts of the submerged rocks and the sandbanks, the chances are that you will run aground, or smash a hole in the hull. You would need to be exceptionally fortunate if, in spite of remaining clueless about where these obstacles truly lie, you nonetheless manage to avoid them all. It is tempting to apply a similar thought to science. How can Newtonian physics put a man on the moon unless Newtonian physics gets things more or less right? Several philosophers have been attracted to the idea that science could not aid us in so many practical ways unless its theories were true, or at least close to the truth.[11] Radically false theories, the thought goes, could yield successes only if their application was supplemented by a prodigious dose of good luck.

The philosopher Hilary Putnam is often credited with being the first to state what is usually known today as the No Miracles argument. It is one of the only arguments that has been put forward in defense of scientific realism. "The positive argument for realism," Putnam wrote, "is that it is the only philosophy that doesn't make the success of science a miracle."[12] In other words, if scientific theories were false, their ability to yield success would be entirely inexplicable, or miraculous. Putnam says that scientific realism—that is, the view that scientific theories are close to the truth—offers us the only tolerable explanation of scientific success.[13]

The problem with the No Miracles argument is that a line of thinking that seems obviously compelling to some commentators—namely, that ideas or hypotheses are unlikely to be useful unless they are also close to the truth—has seemed just as obviously mistaken to others. Writing toward the end of the nineteenth century, Friedrich Nietzsche thought it evident that just because ideas are helpful, this tells us nothing whatsoever about whether they tell us how things are. Mistakes, too, can be beneficial:

> *Origin of knowledge:* Over immense periods of time the intellect produced nothing but errors. A few of these proved to be useful and helped to preserve the species: those who hit upon or inherited these had better luck in their struggle for themselves and their progeny. Such erroneous articles of faith, which were continually inherited, until they became almost part of the basic endowment of the species, included the following: that there are enduring things; that there are equal things; that there are things, substances, bodies; that a thing is what it appears to be; that our will is free; that what is good for me is also good in itself.[14]

Nietzsche's overt comments about Darwin were all negative, but here Nietzsche is suggesting that once we embrace a Darwinian explanation for the preservation of beneficial ideas, there is no longer any need to appeal to truth as an explanation of their success. If hypotheses weren't useful, says Nietzsche, then they would not have survived. That is all the explanation we need for why many of our ideas have such practical value, and it leaves open the question of whether the ideas we have retained reflect how things really are, or whether they are simply errors that work.

The same Darwinian style of explanation has continued to appeal to more recent opponents of scientific realism. Bas van Fraassen, perhaps the most prominent and most thoughtful of modern scientific anti-realists, explicitly counters the No Miracles argument thus:

> I claim that the success of current scientific theories is no miracle. It is not even surprising to the scientific (Darwinist) mind. For any scientific theory is born into a life of fierce competition, a jungle red in tooth and claw. Only the successful theories survive—the ones which in fact latched onto the actual regularities in nature.[15]

Good theories must capture patterns in those parts of the world we can observe, says van Fraassen. It is no surprise that they succeed in doing so, for if they failed, we would have rejected them. A theory's success in predicting what we can observe consequently tells us nothing about whether it truly describes the deep workings of the universe. Success, says van Fraassen, tells us nothing about whether science accurately describes things that are unobservable.

Miracles and Medical Testing

How, in the face of this standoff, can we make progress in our assessment of the No Miracles argument for scientific realism?

An intriguing advancement in recent years suggests that the argument contains an error of probabilistic reasoning known as the "base-rate fallacy."[16] What is this fallacy? It is best to explain it in a domain far removed from the scientific realism debate.

Medical tests can go wrong in two ways. Consider tests for a particular kind of cancer. One test may give a positive result whenever patients really have cancer, while also giving erroneous positive results for many patients who do not have cancer. This kind of test has a high rate of *false positives*. A different test might give negative results whenever patients are free from cancer, while also giving frequent negative results when patients do have cancer. This kind of test has a high *false negative* rate. In real life, medical tests are never perfect, and their designers face a trade-off when it comes balancing the risk of false positives—which can lead to unnecessary worry, and even unnecessary treatment—against the risk of false negatives, which can result in the neglect of serious diseases.

Let's imagine a test that has been developed for the imaginary disease *philosophomania*. Suppose you know that the test has a 10 percent rate of false positives and a 20 percent rate of false negatives. Finally, suppose you test positive. What is the probability that you actually have philosophomania? Some people are tempted to say that they have a 90 percent chance of truly having philosophomania, on the grounds that the false positive rate is 10 percent. But this would be a mistake: in fact, neither the false positive rate nor the false negative rate, nor even the two in combination, contains enough information to allow you to answer the question posed. You also need to know whether the disease in question is common, or rare, in the population as a whole.

To see why this additional information makes a difference, imagine that philosophomania is exceptionally rare. Suppose that in a population of 100 million people, we should expect only about 10 people to have it. Our information about the test's false negative rate tells us that of the 10 or so people who really do have the disease, 2 of them will have a false negative result and the remaining 8 will test positive. Meanwhile, our information about the test's false positive rate tells us that of the 99,999,990 who are likely to be free from the disease, one in every ten will have a false positive result. So in our population of 100 million, 8 disease sufferers will test positive and 9,999,999 people without the disease will also test positive. If, then, you test positive, the chances are overwhelming that you are in the massive group of people free from the disease, rather than in the comparatively tiny group that have it. Your chances of having the disease are not 90 percent: they are roughly one in a million.

The moral of this mathematical story is that you simply cannot make good probabilistic inferences about the significance of medical tests unless you know facts about how rare, or how common, diseases are in populations. These facts are known as "base rates." Base rates are regularly neglected by people who should know better, including elite medical students.[17] But what does any of this have to do with the No Miracles argument?

The proponent of the No Miracles argument tells us that the success of a theory means that the theory is also overwhelmingly likely to be true. This seems to be based on a conviction that a false theory is unlikely to yield success. It may also be based on a conviction that a true theory is unlikely to bring about failure. Perhaps these convictions are reasonable. But we also need to remember our base rates. Suppose that when we survey the entire population of theories, true and false, we discover that true theories are very rare, while false theories are very common. It will then be likely that a successful theory will turn out to be a false one. For even if only a tiny percentage of false theories are successful, and even if most true theories are successful, the fact that false theories are so much more numerous than true ones has the result that most of the successful theories are likely to be false. The No Miracles argument is incomplete unless we know whether true theories are rare among the population of theories considered as a whole, or whether they are instead common.

In summary, the statistical challenge to the No Miracles argument tells us that the argument's defenders have failed to supply crucial information about base rates, which is necessary for their argument to be made good. Moving back briefly to the disease case, it is fairly clear what we mean when we talk about the incidence of philosophomania in the population at large. We are asking whether it affects only a few people, or very many. But it is entirely unclear what we mean when we ask whether truth is common or rare among theories in general. How do we count how many distinct theories there are? Should we think about theories people have formulated, theories they are likely to formulate, or theories that might be formulated even if no one will ever think of them? It seems the No Miracles argument requires that these questions be answered if it is to have any force. But it also seems that these questions are barely intelligible. It seems that the No Miracles argument has fallen apart.

The Suspicious Case of Philosophical Evidence

There is something else that is fishy about the No Miracles argument. Suppose we ask the question "Does DNA have a double-helical structure?" The best way to answer this question involves pointing to all of the evidence amassed—from pictures of DNA produced by the technique of X-ray crystallography, from determinations of the relative proportions of nucleic acids in the molecule, from considerations of the functional role DNA is supposed to have in the workings of chromosomes, and so forth—and determining whether this evidence is better accounted for by the double-helix hypothesis than by other suggestions put forward for the molecule's structure. In other words, the best way to decide on whether DNA has a double-helical structure is to go through a set of processes very similar to those James Watson and Francis Crick (and Rosalind Franklin, and many others) went through when they asked the question in the first place.

The No Miracles argument seems to promise us some further philosophical evidence, beyond these scientific considerations. An additional reason for thinking that DNA has a double-helical structure lies in the fact that the successes of the double-helix hypothesis in accounting for the evidence are best explained by the truth of the double-helix hypothesis.

We should be suspicious of this idea that there is philosophical evidence supporting the double-helix hypothesis, in addition to the basic scientific evidence uncovered by Franklin, Watson, Crick, and others. Imagine that a detective tells his audience, reasonably enough, that he thinks the butler did it because this would give a better account than any other hypothesis for the weapon found in the butler's quarters, the blood on his shirt, and the strands of the butler's hair found on Lord Ashwater's body. The detective then tells us he has additional evidence, drawn from his reading of scientific realist philosophers, that strengthens his case against the butler. He tells us that if his hypothesis were true, this would explain why it accounts successfully for all his observations. Nothing other than the truth of his hypothesis can account for its explanatory successes.

It is clear what has gone wrong here. When the detective says that the truth of his hypothesis explains its success, this is just a way of summarizing, in a compact form, what he has already told us: namely, that if the butler did it, this would offer a better account than any other hypothesis the detective can think of for the location of the weapon, the bloodstains, and the hair strands. The detective does not make a mistake when he says that truth explains success. He makes a mistake only when he suggests that this constitutes an additional piece of evidence for the butler's guilt, over and above what he initially told us. His mistake is to engage in double-counting.

There is no difference between saying "The butler did it" and saying "The butler really did it," or "It's true that the butler did it," or "It's a fact that the butler did it," or "The view that the Butler did it accurately reflects the world as it is," beyond the addition of emphasis, or exhortation.[18] This means there is also no difference between saying "DNA has a double-helical structure" and saying "It is true that DNA has a double-helical structure," or "The hypothesis that DNA has a double-helical structure accurately reflects the world as it is."

This understanding of truth allows us to formulate a version of the No Miracles argument that avoids the problem of the base-rate fallacy and the problem of double-counting. The scientific realist is quite right in saying that the truth of scientific hypotheses is the best explanation for their successes. We now see that this is just a way of stating, in a very general sort of way, that the double-helical structure of DNA is the best explanation for the evidence collected by Watson, Crick, and others; that the pattern of descent from a common ancestor is the best explanation for the evidence collected by Darwin and others; that the nature of the Higgs Boson is the best explanation for the evidence recently collected at CERN; that having molecules made from one oxygen atom and two hydrogen atoms is the best explanation for the known properties of water; and so on, for other leading scientific hypotheses.

Understood in this way, the No Miracles argument is not an incompetent effort at statistical inference, so it is not threatened by the base-rate fallacy. It does not aim to provide additional philosophical evidence that goes beyond what is contained in the evidential cases offered by scientists, so it is immune from the problem of double-counting. Instead, it is a way of expressing a general pattern of deference to that scientific evidence.

If we say that something other than their truth explains the successes of these theories about the constitution of water, the structure of DNA, and so forth, we are saying, in effect, that water might well have some molecular structure wholly different from H_2O, and that this (unstated, merely possible) alternative structure might act in a way that emulates the very evidence that the H_2O hypothesis accounts for. That is how the H_2O hypothesis could be both false and successful. It turns out, then, that the only way to undermine our reformulated version of the No Miracles argument must draw on a general appeal to underdetermination. In other words, the opponent of the No Miracles argument must gesture toward the mere possibility that our best theories might face radically different competitors that can offer equally good accounts of all the evidence. We have already seen that this sort of appeal is dubious. In setting aside the challenge of underdetermination, the No Miracles argument is vindicated.

The Pessimistic Induction

One of the most striking arguments against scientific realism—often associated with the philosopher Larry Laudan—trades on a historical image of science as a series of heroic failures.[19] Time and again, scientists have been more or less certain that they have gotten things right. And time and again, that certainty has been overturned by revolutionary theorizing, or by revolutionary experimentation. Newtonian physics, for example, was thought for centuries to be so secure that some—most notably the philosopher Immanuel Kant—thought not only that it was right but that it was the only *possible* physics. That confidence was proven unjustified when the broadly Newtonian image of the cosmos was replaced with an Einsteinian picture.

Biology, too, seems to demonstrate these radical shifts. Naturalists' observations of broad stability in animal and plant species meant that for a long period of time these species were thought to be immutable. But by the end of the nineteenth century, naturalists thought all species were descended from a small number of common ancestors that, in turn, had been subjected to manifold transformations over immense periods of time. Darwin's own victory may itself turn out to be short-lived. The image we have inherited from him, of a vast tree of nature with many branches springing from a single trunk, has itself been challenged in more recent years by the discovery of what is sometimes call "horizontal gene transfer" or "lateral gene transfer."

A traditional view says that organisms can acquire genes only "vertically"; that is to say, they can acquire genes only from their parents, through the process of reproduction. In the case of humans, for example, we take the view that the genes we have come solely from our mothers and our fathers. We tend to assume that individual humans cannot acquire genes directly from unrelated friends, and that we certainly cannot acquire them from members of entirely different species. But we have known for several years that the inheritance of genes is not always "vertical." Bacteria, among other organisms, can also receive genes "horizontally." This can happen through a variety of mechanisms. For example, viruses can move small elements of genomes from one bacterium to another. The result is that distantly related bacteria can exchange genes with each other.

There are now indications that the phenomena of horizontal gene transfer are not confined to the microbial world. We have evidence that the genomes of several complex multicellular organisms—including worms and insects—have also acquired genes directly from bacteria, and that these processes may be responsible for the acquisition of important adaptations in the animals in question.[20] One study from 2008 has suggested that three different species of fish—herring, smelt, and sea raven—may have acquired their shared ability to produce a natural form of antifreeze via horizontal gene transfer.[21] The upshot of all this is that apparently distinct twigs on the tree of life are in fact in genetic communion with each other, suggesting to many biological commentators that Darwin's image of life's history as a tree whose branches never fuse, but spread ever outward, now needs to be revised.[22]

It might seem as though these repeated revolutions are a testimony to the power of science to show us the way out of our dogmatic slumbers, and toward a proper understanding of the complexities of our universe. But that is not the way that some scientific anti-realists see things. If every time you buy a new toaster it breaks down after six months, then you should probably bet on your next toaster not lasting through the year, either. Similarly, if every scientific theory—even those backed by generations of learned supporters—is shown by later scientists to be mistaken, then we should think our own best theories are most likely mistaken, too. Perhaps the current generation of scientists will be outraged by this thought, but earlier generations of scientists would have been no less outraged at the idea that Newton would be consigned to the wastebasket. Just as Newton once gave way to Einstein, so Einstein's notions of relativity will ultimately be shown to be mistaken. Just as Darwin persuaded the natural historical elite that they were mistaken to think of species as eternal and unchanging, so Darwin's vision of life's tree will also be rejected.

This line of anti-realist argument tells us that the scientific realist does not have history on her side, for good historical reasoning, based on an extrapolation from the failures of past science, should persuade us that today's science will eventually be exposed as erroneous, too. On this account, science does not give us increasingly accurate representations of the universe. It simply trades in one set of productive mistakes for a newer set of productive mistakes. Little wonder this argument is usually called the Pessimistic Induction.[23]

Reasons to Be Cheerful

My colleague Peter Lipton always thought there was something suspicious about the Pessimistic Induction, especially about the way the argument uses historical evidence.[24] In a well-designed experiment, we arrange things so that the evidence we collect allows us to discriminate between the hypotheses we are testing. If smoking does cause cancer, there should be a fairly good correlation between smoking and the disease; if smoking does not cause cancer, there should be a poor correlation. So we can examine whether there is a correlation in order to help us decide whether smoking is, or is not, a cause of cancer. In the debate over scientific realism we are comparing, on the one hand, the realist hypothesis that the history of science is a history in which increasingly accurate images of nature are produced and, on the other hand, the alternative hypothesis that the history of science is a history in which one set of errors is replaced by another set of errors, where no increase in accuracy occurs.

The historical record tells us that scientists repeatedly reject earlier theories and replace them with new ones that are different. But that evidence seems consistent with both hypotheses: it fails to discriminate between them. For even if the scientific realist is right—even if science does produce increasingly accurate images of nature—we should still expect that earlier theories will be discarded as the mistakes they contain are recognized and corrected. This suggests that for the Pessimistic Induction to work as an argument against scientific realism, the history of science would need to demonstrate not just a pattern of replacement and revision but a pattern of regular, wholesale upheaval. A historical signal like that would make it difficult to argue that later theories provide refinements of, or elaborations on, the insights contained in earlier theories. Lipton argued that scientific realism was undamaged by the historical record, because it is far from clear that the history of science provides the evidence needed for these more threatening forms of pessimism.

First, there is straightforward continuity in many areas of science. The periodic table has been added to, but its main groups have remained largely stable for nearly 150 years. Understandings of many of the properties of the major chemical elements have been constant for similar periods of time. As we have seen, recent work in microbiology tells us that elements of the genomes of animal and plant species may have been borrowed from distantly related groups of bacteria. Scientists are also increasingly skeptical of the idea that bacteria themselves stand in simple genealogical relationships with each other. Even so, this has not fundamentally challenged the notion that tree-like diagrams are reasonably faithful ways of depicting the basic evolutionary histories of, for example, animal species. In this respect, Darwin's view is intact.

Second, even when science is accompanied by significant theoretical upheavals, this need not mean that old views are consigned to the waste bin when new ones take hold.[25] It is often pointed out that Newtonian physics gives us a good enough approximation of the behavior of objects traveling below the speed of light for NASA to have used it to send people to the moon. And Darwin did not reject the painstaking work of his anti-evolutionary scientific predecessors when he constructed his theory of evolution by natural selection: instead, he took research that showed deep similarities in the anatomical structures of distinct species, and he used that work to favor his own evolutionary views. This older work was endorsed and given a new interpretation.

Modern work in molecular genetics, which tells us that genes are located in chromosomes, and which identifies genes with stretches of DNA, could not possibly have been anticipated by Gregor Mendel. Evidently a nineteenth-century monk could know nothing of the molecular composition of inherited biological material. Even so, this far more recent work helps to explain why Mendel was able to observe characteristic patterns of inheritance in pea plants, why it was reasonable for him to attribute those patterns of inheritance to the transmission of then-unknown "factors," why Mendel's "laws" of inheritance work when they do, and why they so often fail. Mendel's work is not destroyed by modern genetics, but it is reinterpreted, revised, and reconfigured.

The scientific anti-realist Kyle Stanford, one of the most persuasive and perspicuous proponents of the Pessimistic Induction, has expressed worries about whether this sort of response from the scientific realist is legitimate.[26] Stanford concedes that there are ways of looking back on what Mendel did and arguing that his insights have been preserved in modern genetics. We no longer think that there are strict "laws" of inheritance, because our understanding of how chromosomes are structured, how multiple genes interact with each other in the developing organism, and how they are broken up and recombined during the processes by which sperm and eggs are formed all lead us to think it quite rare that the traits of mature organisms will be transmitted in simple ways across generations. But we can offer an interpretation of Mendel that casts his views as recognizable ancestors of our own, and which tells us that something of those early views has been preserved in the understanding we now have of inheritance. We still think that some traits—especially a small number of major diseases—are sometimes inherited in "Mendelian" fashion, and we think it significant that Mendel also understood this.

The problem that Stanford detects with this method of salvaging continuity between Mendel's work and our own is that the method is entirely retrospective: it is a story told in hindsight. Stanford challenges the scientific realist to give a prospective recipe that will allow us to say which elements of modern theories will be preserved in the science of the future and which will be discarded. If we cannot do this, then while we might be confident that scientists of the future will look back on our own best theories and console us with the thought that we did a reasonable job, we can have no idea *which* elements of our modern views will be preserved and *which* will be consigned to the pyre of embarrassing mistakes. He argues that this challenge cannot be met, and so scientific realism perishes.

I agree that Stanford's challenge cannot be met, but I deny that the realist needs to answer it.[27] If there were a way to tell in advance which elements of current science will end up being preserved, and which rejected, then we philosophers would have found a magical way to press the "fast-forward" button on scientific inquiry. It takes hard work for scientists to discover which of their theoretical commitments are worth hanging on to and which can be jettisoned. Future scientists will know more than we do, and that will include knowledge of where we have gotten things right and where we have made mistakes. The preservation of scientific insight can be judged only retrospectively, but that should not be thought of as a problem for the scientific realist. Nothing else is possible, but nothing more is required, if we are to sustain an image of science as a provider of increasingly accurate representations of the parts of the universe with which it deals.

Further Reading

The best overview of the scientific realism debate (and also a significant defense of scientific realism) is:

Stathis Psillos, *Scientific Realism: How Science Tracks Truth* (London: Routledge, 1999).

Several important articles on scientific realism are collected in:

David Papineau, ed., *The Philosophy of Science* (Oxford: Oxford University Press, 1996).

Probably the most important and influential statement of scientific anti-realism in the past fifty years comes from Bas van Fraassen:

Bas van Fraassen, *The Scientific Image* (Oxford: Clarendon Press, 1980).

Hasok Chang's wonderful study of water raises many significant challenges for the form of realism discussed in this chapter:

Hasok Chang, *Is Water H_2O?* (Dordrecht, Holland: Springer, 2012).

Finally, Kyle Stanford's reinterpretation of the Pessimistic Induction is lucid and engaging:

Kyle Stanford, *Exceeding Our Grasp: Science, History and the Problem of Unconceived Alternatives* (Oxford: Oxford University Press, 2006).

What Science Means for Us

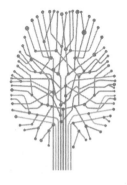

Chapter Five

Value and Veracity

The Division of Advisory Labor

In 2012, the Royal Society—one of the most prestigious scientific academies in the world—together with the Royal Academy of Engineering produced a scientific review of "hydraulic fracturing," a technique for the extraction of shale gas that is more normally known under the notorious name of "fracking." The review had been requested by the United Kingdom Government's Chief Scientific Advisor, Sir John Beddington (himself a fellow of the Royal Society). In the opening sections of the review, the report's authors made their responsibilities clear:

> This report has not attempted to determine whether shale gas extraction should go ahead. This remains the responsibility of the Government. This report has analysed the technical aspects of the environmental, health and safety risks associated with shale gas extraction to inform decision making.[1]

There is an implied division of labor here, typical of reports of this sort, between stating the evidence and offering policy recommendations. In February 2011 the UK Secretary of State for Health asked the Human Fertilisation and Embryology Authority (HFEA) to carry out a similar "scientific review"—this time, to determine "expert views on the effectiveness and safety of mitochondrial transfer."[2] Mitochondria are structures inside animal cells, located outside the nucleus, which contain a very small number of genes that are essential for healthy development and functioning. Disorders of the mitochondria can be systemic and progressive, and they are often passed from mothers to their children. The HFEA was being asked to provide a strictly technical evaluation of a set of novel techniques that hold the promise of allowing people with diseases of the mitochondria to have children who are genetically related to them, and who are also free from these serious diseases. These technical issues were again thought to be distinct from more value-laden concerns about whether it would be right for people to be born—as they would under the proposed techniques—with genetic material from three different contributors, and whether it would be right for fertility clinics to intervene in the human germ-line.

This common division of labor might simply reflect a difference in democratic responsibility: scientists have not been elected, hence it is not their job to say how policy should be formed, even if they have strong views on the matter, and even if the verdict of the best scientific work points clearly in a particular direction. But the division may also suggest to many that there is a strict contrast between the wholly neutral presentation of evidence that derives from science and the evaluative responses various interested parties may have to the evidence. Science, so the story goes, is entirely value-free (or at least, it is value-free when it has not been hijacked by tendentious interest groups). Policy, on the other hand, is what emerges when elected representatives bring their divergent values into contact with objective scientific evidence.

This image of science as value-free might seem to be intimately tied to the scientific realism that was defended in the previous chapter. I defined scientific realism as the view that science provides increasingly accurate representations of the portions of the world it deals with. If science can acquaint us with the facts, it might seem that science must be free of values. For surely there is a distinction between matters of fact and matters of value. The first concern how things are, the second concern how they should be. On this sensible-sounding view, while science tells us how things are, we need to use other forms of reflection, coupled with emotional appraisal, to tell us how they ought to change or whether they should stay the same.

In this chapter we will see that although these linkages between scientific realism and the conception of science as value-free are seductive, they are misleading. Science is permeated with evaluative concerns, but this does not undermine the ability of scientists to reveal the workings of the world to us, nor does it undermine the ability of scientists to advise policymakers on wise courses of action. If science were not informed by values, then the ability of scientists to give prudent advice would be severely limited.

Stalinist Biology

In some notorious cases it seems clear that values have influenced scientific theorizing in ways that are profound and detrimental. The fate of genetics in Stalin's Soviet Union perhaps constitutes the best-known case of all. On July 31, 1948, the

Soviet biologist Trofim Lysenko gave a speech to the All-Union Lenin Academy of Agricultural Sciences in Moscow, during which he reported on the state of biological research. Lysenko's report had been commissioned by Stalin, and Stalin later gave the speech his official approval. Lysenko claimed that the theory of genetics and evolution favored by most American and European scientists was a corruption of Charles Darwin's important work. This genetic theory, which he sometimes referred to as "Neo-Darwinism" and sometimes as "Mendelism-Morganism," was not genuine science at all. Instead, said Lysenko, it was a piece of idealism, or metaphysics.[3]

Lysenko argued that a faulty piece of bourgeois economic theory—namely, the idea that humans, animals, and plants are all locked in a competitive struggle for existence with their fellow species members—had had an unfortunate influence on Darwin, and that its pernicious effects had been magnified by the work of twentieth-century Darwinian thinkers. He went on to claim that the notion of the gene as the persisting, unchanging unit of inheritance—an idea that Lysenko associated with the Austro-Hungarian naturalist and abbot Gregor Mendel, and with the American pioneer of fruit-fly genetics Thomas Hunt Morgan—was an absurdity. It was a manifest fiction that flew in the face of what Lysenko took to be obvious truths about the ways in which the environment could influence organic inheritance, and the ways in which traits acquired during the lives of parents could be passed on to their offspring.

The supposedly idealistic theory of Mendelism-Morganism could not compete with the "creative Soviet Darwinism" that Lysenko championed. This was a "materialist and dialectical approach"—in other words, a properly Marxist approach—which paid due attention to the biological facts, which was oriented toward the practical goal of increasing agricultural productivity, and which accepted that an organism's environmental conditions could be skilfully manipulated so that valuable new capacities would appear in plants and animals. Lysenko called this theory "Michurinism," which he named after the Russian plant breeder Ivan V. Michurin.

Lysenko himself was the son of a peasant. What technical training he had was free of the taint of the prerevolutionary bourgeoisie; indeed, he had very little formal education at all. This made him a suitable emblem for Stalin's own image of the engines of progress. Lysenko's reputation was built on a series of breathtaking claims for his abilities to promote agricultural yields, backed by dubious experiments that he ensured were rarely challenged. Once his brand of anti-Mendelian biology took hold as official Soviet science, the views of Mendelians were denounced as bourgeois, or fascist. This did long-lasting damage to science in the Soviet Union. As the historian Robert Young recalled:

When I was in the Soviet Union in 1971, I met a number of refugees from biology who had found a haven in the history of science. They described the worst effects of shambolic curricula and of censorship in scientific publishing. There were no genetics textbooks published between 1938 and the early 1960s, and no genetics at all was taught to generations of medical students. Imagine trying to practice modern medicine with that gap in one's knowledge. One form of "stupidity" in the period was the inability to memorize and regurgitate Lysenkoist nonsense. I remember one vivid account of a biologist who failed his exams on this topic. On the other hand, there were holes in the net. The original Watson-Crick article on DNA did get published in an obscure work on nucleotide chemistry—which immediately sold out.[4]

Young's comments understate the harm Lysenko did to scientific life in the Soviet Union: scientists lost their jobs, and some died, for their opposition to Lysenko's views. The geneticist Nikolai Vavilov, for example, who had studied in 1913–1914 with William Bateson—one of the earliest pioneers of Mendelian genetics—repeatedly criticized Lysenko's scientific claims.[5] He was arrested in 1940 and died in prison of malnutrition in 1943.[6]

The Lysenko affair shows some of the dangers of mixing science and values. It would be tempting to extend this trite observation in two more general ways. First, one might suggest that good science must be purged of all that is political, ideological, or evaluative. The evidence must simply be allowed to speak for itself. Second, one might conjecture that the Lysenko affair is surely a rare blemish in the history of science—a tyrant such as Stalin was required to sustain such an episode of institutionalized wishful thinking. These days, the thought might go, our scientists are unencumbered by bias. Both thoughts are misplaced, as the rest of this chapter shows.

Women's Orgasms

Hearts are clearly for pumping blood, lungs are for drawing air into the body. But sometimes scientists are unsure of the biological functions of anatomical structures, especially when those structures belong to species that are long extinct. Many species of hadrosaur, also known as duck-billed dinosaurs, had large hollow crests on the tops of their heads. What were these for? Suggestions have included a form of snorkel, an air-tank to enable underwater exploration, and a resonating chamber to amplify calls.[7] We should not suppose, though, that every biological structure must have its own function, as though organisms were composed of neatly designed interlocking elements.

What are the nipples of male humans for? The most tempting answer is that they have no function at all. Male nipples play no role in the survival and reproduction of men. Female nipples, on the other hand, play an obvious biological role in lactation. Although some genes are specific to men, and others are specific to women, the great majority of the genes that are involved in development from egg to adult are common to both sexes. Males have nipples because males and females develop through broadly similar processes, and females need nipples to nourish their young. Male nipples are an evolutionary side effect of female lactation.

What about women's orgasms? What are they for? In a wonderful case study, the philosopher of science Elisabeth Lloyd argues that various forms of bias have affected scientists' work in this domain.[8] Lloyd is happy to acknowledge that the pleasure women get from sex has the biological function of encouraging sexual activity, and thereby reproduction. Her target is instead the specific functionality claimed for orgasm, rather than that claimed for sexual pleasure in general. Lloyd argues that the most plausible hypothesis for female orgasms is that they, like male nipples, have no function with respect to survival and reproduction. Instead, they are best thought of as evolutionary side effects—this time, of the physiological structures underpinning male orgasms. Lloyd is open to the idea that data might eventually be produced demonstrating that women's orgasms do have a biological function. Her claim is merely that as things stand (or rather, as things stood back in 2005 when her book was published), evidence favors the "side-effect hypothesis."

In endorsing what I am here calling the side-effect hypothesis, Lloyd is not asserting that women's orgasms are unimportant, or imaginary, or only mildly enjoyable. Some commentators have attacked Lloyd on the grounds that her skepticism about the biological function of female orgasms devalues them.[9] These attacks are unfair. The abilities to play the piano, to solve complex equations, and to write prose are also unlikely to have functions with respect to survival and reproduction, but there is nothing unreal or frivolous about them. Someone who suggests that sprinting ability, but not footballing skills, assisted the survival and reproduction of our ancestors does not thereby imply that Usain Bolt is a more significant sportsman than Lionel Messi. In order to draw attention to the fact that she regards orgasms as real and valuable, Lloyd has largely dropped her original language of female orgasms as "by-products." That evoked unfortunate images of industrial waste or jars of Marmite. Instead, she now tends to refer to the female orgasm as a "fantastic bonus."

It is not possible to summarize all of Lloyd's evidence in favor of the side-effect hypothesis here, but we can get a flavor of it. Her basic case draws on the facts that, for women, sexual intercourse is often not accompanied by orgasm (even though the women in question are entirely capable of having orgasms) and that orgasms are instead most readily produced by masturbation. This means that female orgasm has no obvious direct link with reproduction. She quotes with approval the American biologist and sexologist Alfred Kinsey's remarks on how intercourse often fails to elicit orgasm: "It is true that the average female responds more slowly than the average male in coitus, but this seems due to the ineffectiveness of the usual coital techniques."[10]

Lloyd goes on to argue that there is little or no credible evidence supporting the various suggestions that have been put forward for biological functions for female orgasms. The zoologist Desmond Morris, for example, suggested back in 1967 that female orgasm helped to solve the potentially fatal problems posed to our bipedal species by gravity. As he put it: "There is . . . a great advantage in any reaction that tends to keep the female horizontal when the male ejaculates and stops copulation. The violent response of female orgasm, leaving the female sexually satiated and exhausted, has precisely this effect."[11] Orgasms tire women out, and cause them to stay lying down. Thanks to this, fertilization is not threatened. A similar hypothesis was put forward in the 1980s, when Gordon Gallup and Susan Suarez suggested that "the average individual requires about five minutes of repose before returning to a normal state after orgasm, and some people even lose consciousness at the point of orgasm."[12]

Lloyd responds by pointing out that the "average individual" Gallup and Suarez specify here turns out not to be a woman at all; instead it is the average *man* who needs five minutes of rest after orgasm, as determined by Kinsey and colleagues in 1948. She also provides evidence indicating that men and women do not respond to orgasms in the same ways: while men might typically need a lie-down, women often continue in a state of arousal after orgasm. Responding to Morris's image of female orgasm keeping the woman prone, Lloyd points out that this presupposes that the orgasmic woman is lying down. She then draws our attention to further research (available when Morris wrote his own piece) indicating that the most effective position for clitoral stimulation, and hence orgasm, during intercourse is when the woman is on top of the man. Under those circumstances, orgasm would seem to encourage, rather than prevent, the draining effects of gravity.[13]

The views of Morris, Gallup, and Suarez are fairly old, and one might think of them as easy targets. Lloyd considers many other theories of the female orgasm, including the far more recent "upsuck" theory, a hypothesis that remains influential today. The basic idea of the upsuck theory is that female orgasm increases the chances of fertilization, because orgasm results in ejaculated sperm being sucked by the uterus into the reproductive tract.

Lloyd recognizes that there is a study, done on just one woman, suggesting that pressure in the uterus drops after orgasm, which might indicate potential for a sort of vacuum suction effect. But she questions the idea that this results in any sperm being sucked into the cervix, or the body of the uterus. For example, she cites a study by Masters and Johnson— pioneers in the 1950s and '60s of the laboratory-based study of intercourse—that reported "[no] evidence of the slightest sucking effect," and she notes that the contractions of the uterus that accompany orgasm may push sperm out rather than sucking it in.[14] She concludes her review with the comment that "three studies suggest no upsuck related to orgasm, and the one study that does consists of a total of two experiments done on the same woman, which document not upsuck itself but a change in uterine pressure."[15]

Although Lloyd claimed there was no good evidence back in 2005 in support of biological functions for female orgasms, she was not foolish enough to suggest that such evidence could never appear. Ten years have passed since her skeptical assessment. Even so, the very best verdict we can come to for proponents of biological functions for female orgasms is that the question remains unsettled.[16] For example, a 2012 review goes against Lloyd's skeptical view, informing readers that "a variety of evidence suggests that female orgasm increases the odds of conception."[17] The authors of that review lean quite heavily on a particular version of the upsuck theory: they claim that orgasm promotes the release of the hormone oxytocin. They also report that, in general, oxytocin promotes the "transport" of sperm through the cervix.

Back in 2005, Lloyd raised an important challenge for this idea: orgasm is not the only way to cause the release of oxytocin, and the amount of oxytocin that orgasm releases is small.[18] Oxytocin levels also increase through sexual stimulation alone, even when orgasm does not occur. The question, then, is whether the boost to oxytocin levels that seems to arise from orgasm is enough to make a significant difference to sperm transport, given that nonorgasmic sexual stimulation appears to raise oxytocin levels all by itself.

Recent work by the sexual physiologist Roy Levin has ended up reinforcing Lloyd's critical treatment of the "upsuck" hypothesis in forceful terms. Levin calls the upsuck theory a "zombie hypothesis"—an idea that simply refuses to lie down even when (from the perspective of the evidence) it is well and truly dead. He notes that the experiments used to show a link between oxytocin release and sperm transport involved injecting women with around four hundred times as much oxytocin as would normally be released in orgasm. So Lloyd's question of whether orgasm releases enough oxytocin to make a difference to sperm transport is a good one.[19] Alongside many other criticisms, Levin also argues that sexual arousal results in the cervix moving into a position well away from the location of ejaculated semen, with the result that even if orgasm produced a suction effect, the cervix would not be close enough to the semen for any of it to be sucked up. His conclusion is blunt: "There is no uncontroversial empirical evidence for the human female's orgasm having any significant role in facilitating sperm uptake by enhancing either its rate or the amount transported or both in natural coitus."[20]

Lloyd concludes, then, that there is no good evidence supporting any story of female orgasm's functionality, and Levin concurs. Why, though, have researchers been so enthusiastic in embracing hypotheses of function, in spite of the poverty of evidence? Lloyd makes two suggestions. First, she suggests there is a bias in favor of *adaptationism*. Very roughly speaking, the adaptationist is one who assumes that the organism can be atomized into distinct traits, each with its own function with respect to survival and reproduction—rather in the manner that an exploded diagram of a washing machine reveals a variety of parts, each of which has a job to do. As we have seen, there is no guarantee that every trait must be explained in this way—it is certainly implausible to think male nipples have biological functions—but researchers on female orgasm seem to have shown a particular enthusiasm for hypotheses framed in terms of biological function, which has led them to overstate evidence in favor of their views, and to overlook evidence against them.

Second, and more interesting, Lloyd suggests that researchers have tended to assume that female sexuality must be like male sexuality: male orgasm has an obvious reproductive function, it is reliably elicited in sexual intercourse, it often results in a period of tiredness. These sorts of assumptions have been projected onto female orgasm in a way that obscures abundant evidence showing how female orgasm and intercourse are only loosely connected. For women, intercourse results in orgasm comparatively rarely, masturbation results in orgasm far more reliably. Indeed, some of Lloyd's earlier work on sex research in primates demonstrates how the presumption that female sexuality *must* be linked closely to reproduction has closed off important areas of research.

Female bonobos (the species formerly known as "pygmy chimpanzees") often engage in something called "genito-genital rubbing": two females hold each other and "swing their hips laterally while keeping the front tips of their vulvae, where the clitorises protrude, in touch with each other." [21] The question of whether this is same-sex *sexual* behavior, or whether instead it is social behavior of a nonsexual kind, seems like a sensible one to ask. But Lloyd points out that this question was closed off from serious inquiry when some researchers stipulated that behavior in nonhuman primates is sexual only when it occurs in oestrus—that is, only when the animal is in a fertile phase of its menstrual cycle and certain hormone measures are high. Since genito-genital rubbing occurs during nonfertile periods, it follows that genito-genital rubbing cannot be sexual. Evidently this is not an important experimental result. It is a trivial consequence of stipulating that behavior can be sexual only if it occurs during a period of fertility.

Darwin's Capitalism

The moral one might draw from Lloyd's work is that various forms of bias distort a true picture of the world. Morris went astray because he assumed, unreflectively, that when women have sex they are like men. Research on bonobos went astray because investigators assumed, without any inquiry, that sexual behavior must be linked to reproduction. These researchers should have set their biases aside and allowed the evidence to speak for itself. On this view, science informed by values is bad science. Good science—science that reveals how things are, as opposed to how we would like them to be, or how we naively expect them to be—is purged of the distorting effects of values.

This conclusion is challenged by the case of Charles Darwin. Darwin is, of course, known today as a natural historian. But Darwin was not a career scientist of the sort who work in laboratories all over the world today. He never held a salaried university position, he did not lecture to undergraduates, he did not chase grant funding. How, then, was Darwin able to fund a lifetime of scientific inquiry? The answer is that he was an exceptionally wealthy man.

Initially Darwin inherited a sizable sum from his father Robert, who (although a medical doctor) made most of his own fortune from investments in canals, roads, and agricultural land. Charles continued this entrepreneurial tradition. His books enjoyed lucrative sales, but the income he received from various forms of speculation, including loans and further investments in land, railways, and the like, far outstripped his earnings from publishing. In short, Darwin was steeped in the industrial capitalist milieu that surrounded the wealthy Victorian entrepreneur.[22]

This capitalist outlook not only funded Darwin's work, it informed it. Darwin's theorizing is saturated with the language of the marketplace, and it is saturated with the vision of agricultural improvement that had helped to make him rich. These aspects of Darwin's writings were noted only a few years after the *Origin of Species* was published. Karl Marx, a great admirer of Darwin, wrote to Friedrich Engels on June 18, 1862: "It is remarkable how Darwin recognises among beasts and plants his English society with its division of labour, competition, opening up of new markets, 'inventions,' and the Malthusian 'struggle for existence.'"

Marx was right about all of this. Darwin frequently used economic forms of argument to suggest that a given biological environment would, over time, contain species that were increasingly specialized and increasingly diverse. Just as economic competition drives traders into new niches, so new ecological niches are opened up by competition in the struggle for life. And just as competition promotes division of labor, so an initially modest stock of biological species can, over time, become diversified into a wonderful array of specialists. For Darwin, nature is a marketplace.

In November 1875, several years after receiving his letter from Marx, Engels wrote his own letter about Darwin to the philosopher Pyotr Lavrov:[23]

> The whole Darwinist teaching of the struggle for existence is simply a transference from society to living nature of Hobbes' doctrine of "bellum omnium contra omnes" [the war of all against all] and of the bourgeois economic doctrine of competition together with Malthus' theory of population. When this conjuror's trick has been performed . . . the same theories are transferred back again from organic nature into history and it is now claimed that their validity as eternal laws of human society has been proved.

Engels' comments differ from Marx's in their tone. Engels seems to suggest that because Darwin's theorizing is a reflection of his Victorian bourgeois economic outlook, this must mean that Darwin's theorizing is unreliable. This was just the line of reasoning that Trofim Lysenko would later endorse, when he claimed that Malthus had led Darwin astray. But why should we accept Engels's inference?

Darwin did indeed see the natural world through capitalist spectacles, but spectacles often help us to see things more clearly. Darwin's theorizing can be shown to be dubious only if we also think that the natural world is nothing like a marketplace. That will take argument; more specifically, it will require that we try to undermine the analogies Darwin draws between competition among members of a species for the resources required for survival and reproduction, and competition among manufacturers for customers.

There are similarities in both domains. In both domains, for example, Darwin suggests that, under suitable circumstances, specialization and increased efficiency can be promoted as though by a "hidden hand": "The more diversified the descendants from any one species become in structure, constitution, and habits, by so much will they be better enabled to seize on many and widely diversified places in the polity of nature, and so be enabled to increase in numbers."[24]

One might try to argue, in a manner reminiscent of Karl Popper, that while the outlook of Victorian capitalism played a role in inspiring Darwin's thoughts, it had no role in the detailed scientific case he made in favor of his vision of evolution by natural selection. This effort to insulate scientific justification from questions of value seems implausible, at least in Darwin's case. We have just seen that Darwin gives us a market-based rationale for how natural selection can promote diversity from initially uniform beginnings, hence why it is reasonable to think that natural selection is the primary agent of nature's spectacular diversity. What is more, the effort to insulate values from the project of scientific justification is unnecessary in any project that aims to vindicate the scientific image of the world: what matters is not, in this case, whether Darwin's views are influenced by his bourgeois ideology but whether that ideology acts to distort, or to reveal, the workings of the natural world.

Sometimes it is capitalism that informs respected theories, but sometimes it is Marxism. Over the past thirty years or so, an important group of evolutionary theorists have begun to stress the ways in which organisms of all types actively construct the environments in which they live. Beavers build dams, which in turn create ponds where beavers are safer from predators and where they have better access to food. Earthworms secrete mucous that coat their tunnel walls, ensuring a damp environment that suits their semi-aquatic physiology. These anecdotes illustrate the foolishness of an image of evolutionary change as a process whereby organisms are the passive victims of active environmental forces. This perspective of "niche construction" has been of considerable value in highlighting the active roles of organisms in determining evolutionary history.[25] And it has its roots in the work of the Harvard biologist Richard Lewontin, a self-confessed Marxist, and a man who explicitly conceived of evolution in Marxist terms as a dialectical interaction between organism and environment.[26]

We must be careful, then, not to generalize from the cases of Darwin and Lysenko to argue that a capitalist approach illuminates nature whereas a Marxist approach distorts it. And we do not have to endorse all—or even many—of the commitments of a capitalist worldview to agree that Darwin's entrepreneurial outlook helped him to see aspects of the natural world that others had missed.

Climate Change and Communication

We have just seen that values play a role as an input to the generation of scientific knowledge. They are also involved on the output side, when scientific knowledge is put to work in the process of policy formation. As usual, this is best illustrated by stepping away from science at first.[27]

Suppose that a friend has come to tea. You serve her a large slice of cake, which you bought from the shops that morning. Before taking a bite she asks, "Are there nuts in this cake?" If her reason for asking is simply that she isn't especially keen on nuts, then you may well reply with a "no," based simply on what the cake tastes like to you. If her reason for asking is that nuts will make her ill, then you might have a fairly close look through the ingredients list before replying "no." And if her reason for asking is that she is likely to suffer a fatal allergic reaction if exposed even to trace amounts of nuts, then you may well take time to study the ingredients list to check if there is a guarantee that the cake is nut-free before replying "no."

In the case of the cake, the amount of evidence you require before responding to your friend with a "no" increases with the costs of error. If you say there are no nuts and the cost of getting it wrong is simply that your friend won't like the cake much, then no great harm has been done and it is reasonable to expend only a little energy gathering evidence for your verdict. If you say there are no nuts and the cost of getting it wrong is your friend's life, then evidently you need to put considerable effort into checking that you are right.

What do cakes and nuts have to do with scientific advice? Suppose a government health official commissions a report on the health risks associated with cell phone use.[28] And suppose a scientist who is compiling the report comes across a poorly designed study indicating that excessive use of cell phones might cause brain damage. Perhaps the study in question has examined a very small number of people who suffered brain damage after using their phones, and it has ignored the need to check these results against the incidence of brain damage among people who never use cell phones.

Should the scientist simply dismiss that study altogether on the grounds that it is methodologically flawed? This would be too quick. The evidence from the study is very weak, but weak evidence should be taken into account under circumstances when the costs of error—in this case, the costs of dismissing a study that might turn out to have uncovered genuine harm— are potentially very high. That is exactly why, if your friend will die from ingesting nuts, you should warn her about nuts in her cake even when you have only a whiff of a reason to think there are any.

Why can't our imaginary scientist, compiling her report for the health official, simply record *all* the available evidence in a way that is uninformed by values? The answer is that a report cannot be infinitely long, and she needs to exercise judgment when deciding what evidence is relevant. Faced with the question of whether to include a poorly designed study, she needs to ask herself about the seriousness of the consequences—that is, she must take a stand on the moral gravity of the consequences—if she dismisses work that is later revealed to have been onto something. It turns out that questions of value are inescapable for responsible scientific activity.

These worries are not merely philosophers' abstractions, cooked up through reflection on an imaginary inquiry into cell phone use. Precisely the same worries have arisen in the context of the reports issued by the Intergovernmental Panel on Climate Change (IPCC), as my colleague Stephen John has recently shown.[29]

Every five years or so, the IPCC produces documents called "Assessment Reports." As the IPCC puts it, the function of these reports is to give policy-makers a summary of "the state of scientific, technical and socio-economic knowledge on climate change, its causes, potential impacts and response strategies." But what sources should be consulted when the sum total of knowledge on these matters is compiled? The IPCC's own answer is that "priority is given to peer-reviewed scientific, technical and socio-economic literature."

Peer-review is a rigorous process of quality control. By requiring that the sources of information for its reports normally be subject to peer-review, the IPCC increases the chances that the work it draws on will be free from falsehoods. That might seem like an unequivocally good thing. But while non-peer-reviewed studies may well contain many falsehoods, they might also contain important truths that, when overlooked, could be disastrous. John illustrates the practical impact of these concerns vividly, by examining the IPCC's changing assessment about the integrity of the West Antarctic Ice Sheet (WAIS). His analysis makes use of important sociological work by Jessica O'Reilly and colleagues, including their interviews with climate scientists.[30]

In its Third Assessment Report, published in 2001, the IPCC raised the possibility that the WAIS might collapse, leading to rising sea levels. But in spite of its acknowledgment that there was "high uncertainty" about the risk of collapse in the long term, the report noted that there was no risk of the ice sheet collapsing before 2100. This consensus had changed dramatically by the time the IPCC's Fourth Assessment Report appeared in 2007. Far from suggesting that the WAIS would remain intact for another century, the Fourth Report suggested that the WAIS might already be in the process of collapsing. In spite of this important acknowledgment, there was no effort to quantify the likely rate of ice loss from the WAIS in either the short or long term, and so the Fourth Report's estimate of future increases in sea levels did not include contributions from the collapsing WAIS.

Why didn't the Fourth Report include a quantified estimate for ice loss from the WAIS? Data and models had been produced well before the Fourth Report's publication that could have produced such estimates, but they had not been published in peer-reviewed form. One scientist complained to O'Reilly and colleagues that "it seemed to us we just couldn't do it [i.e., provide a quantified estimate for the effect of the WAIS collapse] because the IPCC depends on using peer-reviewed results." Of course, if the IPCC's reports began to include results that had not been subjected to peer-review, then the chances the reports will include errors would increase. But the costs of admitting error need to be traded off against the benefits of incorporating valuable work more quickly. The IPCC's reports cannot, and should not, be wholly free of values, because the IPCC must make an evaluative decision about how this balancing act is to be achieved. This statement is not meant to suggest that the IPCC's reports are improper, or unfairly biased: rather, it is simply a statement of the practical necessity of making a value-based judgement about whether to admit evidence that is shaky, but potentially significant.

Taking Sensible Precautions

These reflections on the costs of error and the benefits of timeliness help to give a firm grounding to the "Precautionary Principle," a principle that has been exceptionally important in environmental policy and health policy in the European Union and beyond.[31] There is no single agreed-upon formulation of the Precautionary Principle, but it is often understood, informally, as the notion that when dealing with potentially serious risks to health or to the environment, it is better to be safe than sorry.

Some commentators have taken the view that the Precautionary Principle is objectionably opposed to technical progress, and that it encourages hysterical regulatory responses to "phantom risks." These hostile reactions are easy to understand if we think the Precautionary Principle tells us that whenever some proposed course of action carries the potential for serious harm—even if there is no strong evidence that it will do so—then that course of action should be prohibited. Formulating the Precautionary Principle in this way would result in the banning of genetically modified (GM) crops even if there is only the shadow of a suspicion that "super weeds" might overrun the world. It would halt medical progress, for scientists can never demonstrate with certainty that new drugs, or new fertility treatments, are safe.

This version of the Precautionary Principle is not, in fact, opposed to technology. Instead, as the American academic lawyer Cass Sunstein (who served as President Obama's regulation tsar between 2009 and 2012) has argued, the real problem with this version of the principle is that it is incoherent.[32] It recommends nothing, either pro- or anti-technology. For suppose we suspect cell phones may cause brain damage, even though we admit there is no good evidence supporting this conjecture. And suppose we also suspect cell phones may prevent deaths from abduction and exposure, by allowing people to call home, even though we admit there is no good evidence supporting this conjecture, either. Precaution tells us we must ban mobiles, and that we must not ban mobiles. Precaution tells us nothing.

Fortunately, we do not need to throw precaution to the wind. One of the most important efforts to state the Precautionary Principle came at the "Earth Summit," held in Rio de Janeiro in 1992. Principle 15 of the Rio Declaration stated: "Where there are threats of serious or irreversible damage, lack of full scientific certainty shall not be used as a reason for postponing cost-effective measures to prevent environmental degradation."[33] This principle does *not* tell us that the mere possibility of disaster is enough to veto a proposed course of action. That is just as well, for possibilities of disaster are easy to come by, and they typically accompany all of our possible choices. Permitting the cultivation of GM crops brings the possibility of takeover by super weeds; halting the cultivation of GM crops brings the possibility of prolonging the harmful effects of drought, which new drought-tolerant varieties might allow us to evade.

To clarify what the Rio Declaration *does* say, imagine once again that I am about to distribute cake, but this time at a party for small children. I vaguely recall that the cake has nuts in it, but I'm not sure, because I've thrown away the box it came in. Suppose I am considering warning the parents present that the cake contains nuts. It would obviously be absurd to insist that I cannot issue this warning until I have established with certainty that there are nuts in the cake. My warning costs very little to issue, it is unlikely to do any harm (except to one or two unlucky children who may needlessly forego a slice of cake that is, in fact, nut-free), and it may avert very serious consequences. The Rio Declaration merely codifies this piece of common sense by saying that lack of scientific certainty should not stand in the way of acting to reduce harm, so long as the actions in question are cost-effective.

Under some circumstances this precautionary position will be pro-technology, not anti-technology. If early indications from a clinical trial seem to indicate massive health benefits, and many lives saved, in the event that a brand-new drug were to replace the standard treatment, then mere lack of certainty about its efficacy should not stand in the way of the new drug being more widely adopted, albeit in a carefully monitored fashion.

It is perhaps better not to think in terms of a "Precautionary Principle"—which might give us a recipe for how to act under circumstances of ignorance—but instead to think in terms of a "precautionary stance"—a posture that acknowledges scientific fallibility, and which is mindful of the costs of making mistakes. The precautionary stance reminds us that our actions should be, so far as is possible, reversible, so that if we learn that we've

made a mistake we can undo, or at least limit, the damage arising from our chosen path. In March 2006, at Northwick Park
Hospital in the United Kingdom, six healthy men's lives were
put at risk through the severe adverse reactions they suffered
in tests on the anti-inflammatory drug TGN1412.[34] Evidently,
it would have been better for the men if there had been longer
intervals between each one's dose. That way, the trial could have
been halted before all the participants were exposed.

The influential sociologist Ulrich Beck has argued, in a dra-
matic fashion, that an ethos of scientific purity can have disastrous consequences if carried over to the practical domain of
policy:[35]

> Scientists insist on the "quality" of their work and keep their
> theoretical and methodological standards high in order to
> assure their careers and material success. . . . The insistence
> that connections are not established may look good for a
> scientist and be praiseworthy in general. When dealing with
> risks, the contrary is the case for the victims; *they multiply the*
> *risks.* . . . To put it bluntly, insisting on the *purity* of the scien-
> tific analysis leads to the pollution and contamination of air,
> foodstuffs, water, soil, plants, animals and people.

Beck tells us that scientists are reluctant to assert causal
linkages between chemicals and health risks unless they are
proven to a high degree of certainty. He also suggests that this
reluctance derives, in part, from those scientists' concerns for
their personal wealth and advancement. That is unnecessarily
inflammatory. There are good reasons for scientists to insist
on solidity in their results. If scientific work is to have a cu-
mulative character—if, that is, later generations are to build
on the work of their predecessors—then it is important that its

foundations are secure. In other words, it is important that the body of accepted scientific wisdom is—as far as is feasible—free from error.

This requirement explains the significant burden of proof required before research is deemed reliable enough to enter the expanding corpus of scientific knowledge. We have seen enough in this chapter to understand that these legitimate scientific concerns over evidential reliability must give way when scientific research is put to work in policy. Governments, and the scientific policy committees that advise them, are not primarily concerned with curating a slowly expanding body of reliable information. Instead, their own immediate concerns lie with the health and safety of their citizens. Here, the requirements of timely action demand that policy-makers sometimes act on the basis of poorly designed studies and flawed pieces of research. Slipshod methods do not inevitably produce misleading results. The precautionary stance asks us to remember this.[36]

Further Reading

For broad overviews of debates about science and value, see the following:

Hugh Lacey, *Is Science Value Free?* (London: Routledge, 1999).

Harold Kincaid, John Dupré and Alison Wylie, eds., *Value-Free Science: Ideals and Illusions* (Oxford: Oxford University Press, 2007).

Many of the arguments of this chapter are inspired by the work of Heather Douglas:

Heather Douglas, *Science, Policy and the Value-Free Ideal* (Pittsburgh: Pittsburgh University Press, 2009).

For further details regarding work on female orgasms, see:

Elisabeth Lloyd, *The Case of the Female Orgasm: Bias in the Study of Evolution* (Cambridge, MA: Harvard University Press, 2005).

Chapter Six

Human Kindness

Scratch an Altruist

Has evolution made us good or bad? The verdict of natural history has seesawed over the years. Charles Darwin is still cast, sometimes, as the originator of a grim conception of our moral psychology. Our minds and bodies are the products of a war of all against all, in which weakness is eliminated and the strong survive. If what makes us tick has been shaped by millions of years of bloody struggle, then one might think the victors—modern people—are those with a steely focus on personal advantage. But this was not Darwin's way. He devoted much of *The Descent of Man* to telling a story of moral progress, whereby a variety of evolutionary processes had given us an instinctive feel for the needs of others, made more sensitive and more effective by intelligent reflection:[1]

> I have so lately endeavoured to shew that the social instincts—
> the prime principle of man's moral constitution—with the aid of
> active intellectual powers and the effects of habit, naturally lead

to the golden rule, "As ye would that men should do to you, do ye to them likewise," and this lies as the foundation of morality.

Darwin did not try to argue that evolution had made us into egoistic monsters. He argued that evolution had stamped Christian ethics into our impressionable brains.

Such was wisdom in the 1870s. By the 1970s, it was not unusual for biologists to offer far more cynical interpretations of human motivation, often justified by appeals to what they took to be a more sophisticated conception of evolution than Darwin's. So, the noted evolutionary theorist Michael Ghiselin opined: "If the hypothesis of natural selection is both sufficient and true, it is impossible for a genuinely disinterested or 'altruistic' behaviour pattern to evolve."[2] Evolution, Ghiselin seemed to suggest, had made us selfish. Of course, that does not mean we will all admit to being selfish, for evolution also seizes on the strategic benefits to be had from manipulation and deception. Hence Ghiselin's acid follow-up: "Scratch an altruist and watch a hypocrite bleed."[3]

In the past few years evolutionists concerned with social behavior have once again become convinced of the degree to which humans are prepared to help others, and they have developed an eclectic set of tools for explaining how evolutionary processes could lead to such tendencies. In the remainder of this chapter we will address some of the confusions that have arisen when scientists have tried to explain the evolutionary origins of moral behavior.

Selfishness and Altruism

The cynic tends toward a miserly interpretation of the behavior of others. When he witnesses what looks like a piece of well-meaning assistance, he writes it off as an effort to impress onlookers with one's largesse. When a noted pop star begins to campaign for global justice, the cynic wonders if the celebrity might not be looking for an opportunity to rub shoulders with the elite in Davos. The cynic's mean-spirited position needs evidence: it is a bold conjecture to suggest that people are fundamentally selfish, and there is nothing immediately plausible about it. Darwin thought it too much of a stretch to reinterpret every action as motivated by self-interest, if only because some actions seem to occur so quickly that they seem incompatible with calculation of any kind:[4]

> Many a civilised man, or even boy, has disregarded the instinct of self-preservation, and plunged at once into a torrent to save a drowning man, though a stranger. Such actions . . . are performed too instantaneously for reflection, or for pleasure or pain to be felt at the time.

Cynicism about human motivation is hardly a default option, so why has it been thought credible?

The answer lies in a seductive, but fallacious, link between an evolutionary image of our species and a cynical image of our human morality. Evolution, one might think, is a matter of struggle between individual organisms. The fitter specimen will tend to prevail. And so, natural selection is bound to favor traits that benefit the organisms that bear them.

This sort of view is encouraged by some of Darwin's own comments. He took it that natural selection "works solely by and for the good of each being."[5] If natural selection cannot favor traits that benefit others, it seems that natural selection cannot favor anything other than selfishness. And so, moving to the early years of our current century, we find the influential sociobiologist Richard Alexander casting doubt on the intentions of apparently ethical investors:[6]

> It is difficult to imagine that anyone invests in the stock market for altruistic reasons; perhaps no one invests in anything at all (except relatives) without expecting (not necessarily consciously) phenotypic rewards that include some kind of interest on the investment.

Alexander's particular brand of Darwinism—very unlike Darwin's own—rules out altruism. Care for one's offspring is acknowledged, but Alexander's evolutionism dictates that it must be understood as a form of selfishness: "Reproduction," he says, "is a selfish act, meaning only that it serves the life interests of the reproducer."

In the remainder of this chapter I want to suggest that the sort of message one might take away from reading Alexander's work is a distortion of the best that biological research has to offer. There is no reason to think that evolutionary study bolsters the cynical vision of human morality, and every reason to think that evolutionary work makes room for all sorts of human kindness.

Two Types of Altruism

Our first step to untangling the confusing proliferation of views about selflessness and cynicism requires that we get our concepts straight. In some sense or another altruism is about benefiting others, while selfishness is about benefiting oneself. But what, exactly, are we committed to when we describe someone as altruistic? We might be making a claim about the person's character. More specifically, we might be talking about the reasons that typically motivate her actions: an altruist is someone who acts because of her concern for the welfare of others, while a selfish person acts because of concern for her own benefit. We can call this the *psychological notion of altruism.*

Note that because this definition is about psychological motivation, it can be applied only to organisms that have psychological states. Bacteria can achieve remarkable things, but they do not do so on the basis of reasons, so it is absurd to ask whether bacteria are psychologically selfish or psychologically altruistic. Bacteria are profoundly blithe: they do not care about anything, including themselves. It is also a feature of this psychological definition that altruism is independent of success: what matters for an assessment of psychological altruism is what causes you to undertake a specific course of action, not whether that action ends up benefiting others. A psychological altruist's plans may go awry, and she may even end up benefiting herself more than others, but that will not suffice to make her psychologically selfish.

This psychological definition, focused on motivational reasons, stands in sharp contrast to the biological understanding of altruism common among evolutionary theorists. Altruism—but only altruism when understood in a particular way—poses a problem for simple understandings of natural selection. For suppose we think of altruism not in terms of the psychological causes that motivate behavior but, instead, in terms of the effects of a piece of behavior on others' survival and reproduction. Consider, for example, the misery of the male emperor penguin in the Antarctic winter. He must survive temperatures that sometimes drop below minus 49°F and winds of up to 110 miles per hour. He must do this while incubating an egg, and without taking on food.

The key to enduring such conditions is to snuggle up close: penguins cuddle up so tight that around twenty-one birds pack into each square metre of frozen ground. Temperatures at the centers of these cozy spaces sometimes get as high as 98.6°F.[7] Of course, things are much colder for the penguins on the outside of the huddle, but by taking turns on the periphery, every penguin can benefit. Suppose, then, that some unscrupulous penguin were to stay stubbornly in the center of the huddle, never enduring the cold of the perimeter. He would receive all the benefits of huddling, but without paying any of the costs. Those who take their turn can be considered altruists relative to such a freeloader: their actions leave him better off than they are.

A *biologically altruistic behavior* is usually understood by evolutionists to be one that augments the ability of others—call them "recipients"—to survive and reproduce, while damaging the survival and reproduction of the organism producing the behavior—call it the "actor." In other words, altruistic behaviors increase the reproductive fitness of recipients while reducing the reproductive fitness of actors. Biological altruism, unlike psychological altruism, has nothing whatsoever to do with character, or motivation. So while there is little point in asking whether bacteria might be psychologically altruistic, it might well make sense to ask whether bacteria are biologically altruistic. In fact, this is no mere possibility. The question of altruism regularly arises in the study of microbiological behavior. While they do not make plans, while they have no personality, bacteria are magnificently social organisms. As my former student Jonathan Birch puts it, "we now realize that what looked like a blob on a Petri dish is in reality a dynamic social network."[8]

Myxococcus xanthus colonies move in ripple-like formation when close to food sources, perhaps in a way that allows prey to be dislodged more effectively. These bacteria may be coordinated pack hunters. Many other bacterial colonies produce chemicals that are released into their environments, and their

effects—whether poisonous, adhesive, digestive, or otherwise—
are to the advantage of all the colony's members. This means we
can ask the same questions of bacteria as we ask of penguins:
since there is a cost, in terms of metabolic effort, to producing
these beneficial chemicals, could we not just as easily imagine
bacterial freeloaders that take advantage of the goods produced
by other members of their colony, without going to the trouble
of contributing themselves? Wouldn't such a bacterium do better than its rivals? Wouldn't it take over the colony, eventually
outcompeting the cooperators that it initially took advantage
of? The question of biological altruism arises for all organisms,
not just for those with brains, and not just for those that act on
reasons.[9]

This distinction between biological and psychological altruism immediately allows us to dilute many of the apparently uncomfortable consequences of evolution for our moral
self-image. Natural selection can favor parents who care for
their offspring: an individual who gives resources to its babies
may have more healthy children and grandchildren to inherit
its beneficent habits than an individual who hogs all available
food and whose babies starve as a result. For that reason, pa-
rental care is usually thought of as biologically selfish. But this
tells us nothing about what psychological states—assuming the
organism in question has psychological states—might motivate
such beneficence.

There is no contradiction in thinking that natural selection
has bestowed organisms with genuinely selfless concerns for the
well-being of their children. To the extent that we are interested
in evaluating people's character, it is matters of psychological
motivation that are of primary interest to us. Richard Alexander's comment that "It is difficult to imagine that anyone invests
in the stock market for altruistic reasons" overlooks the obvious

range of cases where people invest in the stock market to secure an inheritance for their children.

Psychological altruism with respect to nonkin can also be favored by the simple action of natural selection on biologically selfish traits. For the purposes of illustration, imagine a society where some people are *miserly,* the rest are *generous of spirit.* A miser cares only for himself; he is never moved by the plight of others, and he keeps his resources squirreled away as a result. The generous of spirit, on the other hand, willingly share what they have with others. They do so because they care for the well-being of their fellows. But let us add that the generous of spirit share only with those who are morally deserving; more particularly, they refuse to share with misers. This means that misers never receive donations from others, while the generous of spirit often do. And that, in turn, leaves misers vulnerable to ill fortune: a lean year for a miser can be fatal, but the generous of spirit enjoy a social safety net. In an environment of chancy harvests we can expect the generous of spirit to live longer, and to have more healthy babies, than the misers. That is true in spite of the fact that our misers are psychologically selfish, while the generous of spirit are psychologically altruistic (albeit rather holier-than-thou).

Robert Trivers, who provided a seminal mathematical exploration of the evolutionary significance of these forms of selective sharing, called this kind of phenomenon "reciprocal altruism."[10] It has often been pointed out that, from a strict biological perspective, "reciprocal altruism" is a misnomer: in our imaginary example above, generosity of spirit is not biologically altruistic at all, because the benefits enjoyed by the generous of spirit leave them fitter over the course of a lifetime than the misers.[11] The generous of spirit turn out to be biologically selfish. We should not suppose, though, that the

psychological motivations that underlie these forms of reciprocity must be selfish, too. Trivers's mechanism illustrates
once again how natural selection can promote genuine forms
of psychological altruism.

Selfish Genes

Richard Alexander is unusual among sociobiologists in being
poorly attuned to the crucial differences between psychological and biological conceptions of altruism. It is certainly not a
mistake Richard Dawkins makes. At the beginning of *The Selfish Gene,* a conspicuous cultural landmark, he is careful to tell
us that in discussing biological conceptions of selfishness and
altruism, "I am not concerned here with the psychology of motives. I am not going to argue about whether people who behave
altruistically are 'really' doing it for secret or subconscious motives."[12] Dawkins's reflections on genetic selfishness do not tell
us in any direct way about moral character.

The question of the value of Dawkins's selfish gene approach
has considerable interest for biological explanation, but it is in
many ways a red herring if we are considering the more general
issue of what evolution has to say about human goodness.[13] The
traits that are of interest to evolutionary biologists are the traits
that are inherited from one generation to the next. These are the
only traits that natural selection can favor, or frown upon. If,
for example, the running speed of a predator is to increase over
evolutionary time, then the offspring of faster runners need to
run faster than average, too.

Biologists usually assume that when traits are inherited, it
is because parents transmit genes to their offspring. For most
of the discussion that follows we will further assume that genes
can explain not only the inheritance of traits like running speed

but also the inheritance of tendencies to benefit others. Toward the end of this chapter, and also in the following chapter, we will see reasons to question the thought that inheritance must be genetic. For the moment, it is important to stress that the assumption of genetic inheritance does not mean that genes alone cause baby predators to turn into fast-running adults, nor does it mean that the causal effects of genes as a baby grows to be an adult are inescapable. The speed at which a predator runs is decided not only by its genes but also by the quality of its diet, its luck in avoiding accidents, and many other nongenetic factors. Mainstream evolutionary theory requires only that genes—alongside many other influences—make reliable differences to how development proceeds from egg to adult.

This rather modest understanding of how evolutionary processes work has the immediate consequence that a trait cannot be successful in a species—that is, it cannot become present in a high proportion of the species' members—without the genes that underpin the inheritance of that trait also enjoying success. This also means that when a biologist thinks about how change might occur in a species over time, the biologist can take what is sometimes called the "gene's-eye perspective." The biologist can ask "What would a gene have to do to succeed in this population?" In other words, the biologist can think of all evolutionary processes in terms of what genes must "try" to do, in order to increase their level of representation over the generations.

Dawkins himself advocated this selfish gene approach as one useful perspective from which to investigate nature, and there is evidence that plenty of biologists do indeed find that way of thinking helpful.[14] But when adopting the gene's-eye perspective we must remember that genes do not literally try to do anything: genes merely have effects, which may be favored in some circumstances and penalized in others. We have seen that genes can sometimes be favored because they make their bearers psychologically altruistic. Since genes have no motives of their own, we cannot use Dawkins's ideas to argue that the apparent selflessness of much human action is a cunning disguise that masks deeper, uglier agendas.

The Reality of Altruism

Natural selection does not rule out psychological altruism, because biologically selfish behaviors can be psychologically altruistic. This means our problem has been alleviated, but it has not disappeared. It might seem to follow as an immediate logical consequence of our definition of biological altruism that natural selection—when understood as a process favoring individual organisms according to their abilities to survive and reproduce—can never promote biologically altruistic behaviors. Does biological thinking have the consequence that all behavior ultimately benefits the actor?

Darwin himself did not write in terms of fitness and altruism, but he did appreciate the problem posed for his theory by behaviors that benefit others. Darwin understood morality in humans to be the result of sympathy: we feel the misery of others as though it were our own, and this gives us a spur to help those in peril or pain. But why, Darwin asked, are we equipped with such a sense of fellow-feeling?

> How, within the limits of the same tribe, did a large number of members first become endowed with these social and moral qualities? It is extremely doubtful whether the offspring of the more sympathetic and benevolent parents, or of those

who were most faithful to their comrades, would be reared in greater numbers than the children of selfish and treacherous parents belonging to the same tribe.[15]

Just like the freeloading penguin, shouldn't natural selection favor an individual who was prone only to act for himself and his progeny? He would gain from the beneficence of others, but he would never pay the cost of helping them.

There is evidence that people regularly act in ways that are of benefit to nonkin, and it is these forms of action that the very simplest forms of natural selection struggle to explain. Consider the "Ultimatum Game," a game for two players of all ages. There is a sum of money—let us suppose it is $10—and player one must choose how much, if any, to offer to the second player. If the second player accepts that allocation, then both keep their respective shares. If the second player refuses, then both get nothing. How much should player one—"the proposer"—offer to player two?

If we are all purely self-interested, and if we know we are all purely self-interested, then player one should always offer a penny to player two and keep the remaining $9.99. After all, player two can either accept the deal, in which case she gets a penny, or she can turn it down, in which case she gets nothing. So player two should accept a penny. Knowing that this will be player two's decision, player one should not offer anything more. When real people play the Ultimatum Game, this is almost never what they do. Proposers from places like the United States and Europe typically offer half of the pot of money to the second player.[16]

One cannot simply dismiss these results as irrational: there is no credible theory of rationality that tells us we should care only about ourselves. Instead, these results suggest that people's conceptions of fairness give them a desire to apportion resources in a reasonably equal way. Curiously enough, there is some evidence to suggest that the activity of studying economics—perhaps because it encourages the thought that self-interest is rational— may increase people's levels of selfishness in games such as these, although it has since been suggested that economics may simply attract more selfish students in the first place.[17]

The unusual behavior of economists hints at variation within cultures. Responses to the Ultimatum Game also vary from one culture to another, suggesting different conceptions of what is a fair offer as well as different understandings of what sorts of offers should be punished. Early work by the evolutionary anthropologist Joseph Henrich, for example, indicated that the Machiguenga of the Peruvian Amazon—people who rarely cooperate with others from outside their extended family— typically offer only 15 percent of the pot in the Ultimatum Game.[18] In spite of all this variation, across very large parts of the world typical offers stay stubbornly at around 50 percent, far higher than pure self-interest would lead us to predict. So there is little evidence that evolution has made most of us selfish, even if this has been evolution's effect on economists.

It is reasonable to cast doubt on how much these very artificial games tell us about behavior in day-to-day situations, where people interact in much richer contexts than the stripped-down environment stipulated by simple games. It is very rare indeed that we would ever be placed in a real situation where a sum of money simply falls from the sky and we can offer some to an anonymous stranger. Instead, when faced with a decision about how much money to allocate to someone else, we are likely to already have, or to want to seek, information about where the money came from. Did I earn it? Did we both earn it? Was it stolen? Was it donated by someone wealthy, or someone poor? Why was it donated? We are also likely to want to know the situation of the people I might share it with. Do I know them? Are they sick? Do they have dependent children? Finally, we might want to know what the repercussions of my choices might be. Will there be any follow-up? Might I risk arrest, or intimidation? Research that tries to make these situations more realistic still delivers verdicts that depart from pure self-interest, and our day-to-day experience also provides ample examples of people acting in ways that benefit others, even when those others are unrelated. Many of us give money to charity, most of us act peaceably toward other members of our communities, most of us pay taxes, few of us steal even when we are confident we could get away with it, most of us are civil to others even when we are unlikely to encounter them again.

Evolutionists have not responded by trying to explain away these data about beneficent behavior: instead, they have developed an arsenal of theoretical resources that greatly extend Darwin's framework for thinking about selection. To understand the development of these approaches it is helpful to consider the solution to the problem of altruism that Darwin himself put forward. Recall that Darwin was worried that when considering individuals in competition with each other, a moral disposition to assist nonkin would seem like a handicap:

> It must not be forgotten that although a high standard of morality gives but a slight or no advantage to each individual man and his children over the other men of the same tribe, yet that an increase in the number of well-endowed men and an advancement in the standard of morality will certainly give an immense advantage to one tribe over another.[19]

While selection acting on individuals will tend to eliminate altruistic behavior, such behavior might be favored if we think of a different form of natural selection acting at the level of what Darwin calls tribes, or communities. This evolutionary mechanism is more usually referred to these days as *group selection.*

Subversion from Within

In the 1960s and '70s, biologists such as George Williams and John Maynard Smith subjected group selection to strong criticisms, with the result that many theorists looked on the process with suspicion. The problem was that the case for group selection was often made in a hand-waving manner. Perhaps, as Darwin suggests, tribes with moral members will do better in war than tribes with wanton members. But should this make us confident that morality will emerge by a process of group selection? Might not moral groups instead be "subverted from within," overrun by idle backsliders, with the result that wars occur between increasingly incompetent and spineless communities? Yes, the advantage will go to the better-organized group, but what confidence should we have that this form of advantage can overwhelm the rot that spreads from inside?

Evolutionary thinking has responded to these concerns by introducing new levels of mathematical rigor in the explanation of social behavior. These more disciplined forms of thinking all point to a recognition that (roughly speaking) altruistic behaviors can evolve, so long as the benefits of altruism fall disproportionately on other altruists. Even more roughly speaking, altruism can evolve if altruists clump together.

This point is easiest to appreciate if we focus on two highly simplified cases that demonstrate the two extremes of altruistic clumping. Imagine, in both cases, that genetic inheritance means that selfless organisms usually have selfless babies, and that selfish organisms usually have selfish babies. Imagine, in addition, that the organisms in question are asexual: an organism can have a baby by itself, with no need for a mate. In the first case, let us imagine that our organisms do not care whom they live with, so long as they stay in the group they were born into. And suppose we start with a few selfish individuals scattered across several groups of selfless organisms. Groups with fewer selfish individuals will do better than groups with many such individuals, but in every group the selfish ones will do better than the selfless ones in that group, for they receive the benefits of selflessness without paying the costs. Selfishness will continue to increase within every group until there are no selfless individuals left. Under these circumstances, subversion from within is fatal to the evolution of altruism.

In the second case, let us suppose that selfless organisms instinctively and reliably seek each other out, and that selfish ones do the same. Groups of selfless individuals will then form up, and they will do much better than groups of selfish individuals. Moreover, subversion from within will not be a problem here: our stipulation that organisms seek out like organisms to live with ensures that if a selfish individual happens to be born into a selfless group, it will migrate away until it finds some other selfish individuals to live with. In the population as a whole—in other words, considering all the groups together—we can expect selfless individuals to increase in numbers until they replace the selfish ones, and we can expect this situation to be stable even when selfish individuals are occasionally born by chance genetic mutation. The moral of the story is that if selfless and selfish individuals are distributed in the right kinds of ways, then altruism can evolve.

What, then, does recent evolutionary work tell us about the legitimacy of appeals to "group selection"? The answer is mixed. As we have seen, altruism can evolve when altruists clump together. This means that the way in which a large population is divided into smaller groups makes a difference to the sorts of traits that natural selection can favor in that larger population. In that sense, group selection has been vindicated. But this way of thinking about "group selection" is demanding: the mere fact that organisms' behaviors are beneficial to the groups they live in does not ensure that these behaviors will be likely to evolve. That is the lesson of the problem of subversion from within.

The primary innovations within evolutionary theorizing that allow us to understand the evolution of altruism can all be thought of as showing, in one way or another, why we should expect the sort of clumping together of altruism that is required for its evolution.[20] Most obviously, genealogical relatedness is a mechanism that allows this. Suppose again that some genes cause their bearers to be selfish, while alternative genes cause their bearers to be selfless. If organisms typically interact with their parents and their siblings, and if processes of inheritance mean that different members of the same family tend to have similar genes, then we have reason to think selfless individuals will clump together. The mechanism of *kin selection* proposed by William D. Hamilton—the mechanism popularized by Richard Dawkins in *The Selfish Gene*—shows how genetic relatedness can explain the evolution of altruism (and the evolution of many other kinds of social behavior) in just this way. But Hamilton himself understood very well that genealogical relatedness was just one means by which like organisms might end up interacting with like.

Hamilton's remarks on the importance of genetic relatedness might seem to imply that what matters for the evolution of altruism is that different organisms are all members of the same family: that is, that they are related by descent. In fact, Hamilton's notion of relatedness is more general, and more technical, than this strictly familial notion. For Hamilton, to say that two organisms are related is simply to say that they share genes in common. Organisms that share genes could indeed end up associating with each other because they belong to the same family. But there are many other mechanisms that might have the same effect: perhaps organisms that share genes actively seek each other out, or perhaps organisms that share genes seek out the same food sources, and they associate with each other as a by-product of this.

This is the moral of Richard Dawkins's hypothetical example of the "green-beard effect," an idea that Dawkins also adapted from earlier work by Hamilton. Suppose there is a gene—call it the "green-beard gene"—with two effects. First, it makes the individuals who have it grow a green beard. Second, it makes those individuals seek out and help others with green beards. Individuals with the green-beard gene will now clump together, and they will help each other. They will do this even if they come from entirely different families. In Hamilton's language, they show a high degree of genetic relatedness, even though they are unrelated in a genealogical sense. The moral of the story is that kin selection does not act only on kin.[21]

Of course, Dawkins introduced the green-beard example as a thought experiment, an imaginary curiosity designed to illustrate a conceptual point. Subsequent research has suggested that nature contains real genes with green-beard-like effects. When red fire ant queens possess a certain gene, it causes them to emit a smell. Other ants with that same gene use this smell to recognize which queens possess it and which do not. They kill queens that do not have the gene, while sparing those that do.[22] In other words, the gene in question produces a recognizable odor, and it also promotes beneficent behavior—in this case, the behavior of sparing rather than killing—toward others with the same gene. The odor the gene produces is the equivalent of a green beard: the gene's possessors recognize each other, and act favorably toward each other, thereby benefiting that gene's future prospects.

Darwin Revived

Theorists are beginning to realize that Hamilton's insights have far more general application than was initially thought. Earlier in this chapter I mentioned the general assumption within evolutionary biology that resemblance between parents and their offspring is secured by the passing on of genes. The question of whether inheritance across the animal and plant kingdoms might sometimes be achieved by the action of additional nongenetic mechanisms is a lively one.[23] Whatever we think the answer to this broad question might be, it is obvious that in our own species important behaviors, practices, and technologies can be sustained across generations not because of the transmission of genes but because we learn from each other.[24] This raises the possibility that individuals with genes that promote altruistic behavior might clump together—that is, Hamilton's "genetic relatedness" might be high—because of cultural forces. Ostracism, other forms of socially enforced conformity, and even deliberate migration into prosperous groups might explain why altruists end up interacting primarily with each other rather than with more selfish individuals.

Even more recent work has tentatively extended the evolutionary role of culture even further. Once again, Hamilton teaches us that altruism can evolve if altruists clump together. Perhaps, then, altruism can evolve not merely when cultural forces explain why altruists spend time with each other but also when learning from others—rather than the passing on of genes—explains why babies grow up to be altruistic in the first place.[25] Children can resemble each other with respect to their moral character because they have been inculcated in the same school ethos or because they try to emulate the same role models, not because they share genetic material. If cultural influence can potentially bring about both the acquisition of altruistic tendencies and the reliable clumping together of altruists, then we have further reasons to doubt that evolutionary processes can favor altruism only when the benefits of altruism fall on the actor's relatives.

Hamilton's basic insight has been generalized and made more complex in many ways in recent years. These developments have moved us far from a naive evolutionism, which tells us that the only circumstances under which we act to benefit others are when those others are our family members. When modern evolutionary theorists come to explain our tendencies to help others, they are no longer confined to explanations that restrict these tendencies to nepotistic forms of assistance, and they are no longer confined to genetic conceptions of how these tendencies are inherited. They have developed a rich theory that takes crucial note of the structure of social groups, of communication, of conscious choice about whom to consort with, and about the roles of learning in moral development. While these frameworks have a level of mathematical formality that would have eluded Darwin, they paint a picture that has much in common with Darwin's eclectic approach to the evolution of our moral tendencies. Modern evolutionary theory rejects the cynical recasting of our beneficent behavior toward others, and it is open to the positive role of culture in explaining why we are prone to assist people we have never encountered before. When the modern evolutionist scratches an altruist, the chances are it is an altruist who bleeds.

Further Reading

An important and accessible discussion of various forms of altruism, and a defense of the role of group selection in bringing them about, is:

E. Sober and D. Wilson, *Unto Others: The Evolution and Psychology of Unselfish Behaviour* (Cambridge, MA: Harvard University Press, 1999).

The debate over group selection is covered in detail in:

Mark Borello, *Evolutionary Restraints: The Contentious History of Group Selection* (Chicago: University of Chicago Press, 2010).

For a gripping historical account of some of the most important innovations in evolutionary approaches to altruism, see:

Oren Harman, *The Price of Altruism: George Price and the Search for the Origins of Human Kindness* (London: W. W. Norton & Co., 2011).

Finally, for an overview of some of the most recent (and contentious) theorizing about altruism, see:

Martin Nowak and Roger Highfield, *SuperCooperators: Altruism, Evolution, and Why We Need Each Other to Succeed* (Edinburgh: Canongate Books, 2012).

Chapter Seven

Nature: Beware!

A Modern Superstition

The impression one gets from the mainstream of popular scientific work is that there is a lively debate going on over how much of human behavior can be attributed to nature and how much to culture, learning, socialization, or a variety of other nurture-like processes. For example, the cognitive scientist Stephen Pinker's book *The Blank Slate: The Modern Denial of Human Nature* intimates through its title that human nature should not be denied, that its description is an important task, and that woolly-headed social scientists are responsible for playing it down.[1]

We also find a notion of human nature put to work by politically conservative thinkers in an effort to cast doubt on the wisdom of various technological innovations, especially in the domain of human reproduction. Should we contemplate using genetic engineering to alter human nature? Should we perhaps have more respect for the integrity of human nature? Michael Sandel, a political philosopher prominent in the public domain, has suggested that efforts to alter human beings through the use of genetic engineering or pharmaceutical augmentation should not "override [a child's] natural capacities but permit them to flourish."[2] That presupposes we can determine which capacities of a child are the natural ones, so that we can tell the difference between distortion and encouragement of what nature bestows.

Leon Kass, the former chairman of President George W. Bush's President's Council on Bioethics, sometimes hints at the importance of respect for human nature, and for the broader nature of mammals in general, in some of his remarks on the ethics of cloning. Cloning, if it were ever permitted, would be an asexual form of reproduction, for only one parent would be required for the generation of a new child. And yet, as Kass puts it: "Sexual reproduction . . . is established . . . by nature; it is the natural way of all mammalian reproduction." Hence cloning "shows itself to be a major alteration, indeed a major violation, of our given nature as embodied, gendered, and engendering beings."[3]

It might come as a surprise, then, to learn that some well-informed thinkers have claimed that the very idea of human nature has no place in the light of recent scientific research. The prominent philosopher of biology David Hull was consistently "suspicious of continued claims about the existence and importance of human nature."[4] The biologist Michael Ghiselin—also noted for his seminal contributions to the history and philosophy of science—has been even more direct: "What does evolution teach us about human nature? It teaches us that human nature is a superstition."[5] If there is no intelligible notion of human nature, then it makes no sense to argue over how much of human behavior and thought is due to nature and how much to culture, and it makes no sense to think that appeals to human nature carry any sort of ethical weight. So how good is the case against nature?

Cultural Variability

A large proportion of the world's psychological research is done in universities in the United States, the United Kingdom, and other wealthy industrialized countries. This means that the students attending these universities are sitting ducks when it comes time to recruit volunteers for studies. The consequence is that we know far more about how this particular type of person thinks than we know about how people think in general. As Joseph Henrich and his collaborators put it, our standard psychological research subjects are from WEIRD (Western, Educated, Industrialized, Rich, Democratic) societies.[6] This would not matter much if extrapolation from these students to people in general had been shown to be legitimate, but it has not. The idea that there are behaviors or modes of thought that are common to all people the world over sometimes gets an unwarranted boost from rash generalizations based on what we know about rich Western students.

The previous chapter softened us up to this finding. There we saw the considerable variation in the type of offer made in the Ultimatum Game from one culture to the next. There are plenty of other cases where the examination of other cultures casts doubt on the propriety of hasty generalization from Western students to all people. For example, for a long time many philosophers thought of our vulnerability to visual illusions as the sort of things that could not be affected by learning or upbringing. Here, too, our confidence is ebbing away. Consider the famous Müller-Lyer illusion:

Most readers of this book are likely to see the top line as longer than the bottom one, even though measurement with a ruler will reveal that they are the same length. But this is just an artifact of this book's likely readership, for not everyone sees the lines in the same way. Henrich and company have re-alerted the cognitive science community to a study conducted by Marshall Segall and others in the 1960s suggesting that San foragers of the Kalahari do not see the Müller-Lyer illusion as an illusion at all. They correctly see the two lines as having the same length. Many other cultures, it seems, show a far less pronounced response than Americans.[7]

Henrich and Co. could have mentioned much earlier work on the Müller-Lyer illusion, published in 1901 by the Cambridge anthropologist and physiologist W.H.R. Rivers.[8] Rivers, too, found the responses to the illusion from the undergraduate students he tested in Cambridge to be more extreme than those from the Murray Islanders he tested during his expedition to the Torres Straits. Segall suggested that vulnerability to the illusion depends on how we are brought up. It is more pronounced for people raised in environments full of straight lines and clean angles. For that reason the American subjects usually recruited to psychological studies show the most extreme vulnerability to the illusion of all the world's people.

It has also been suggested that enculturation may affect our ability to tell different colors apart.[9] Russian speakers have no generic term for "blue," which covers all the shades classified as "blue" by speakers of English. Instead, they have two wholly different terms—*goluboy* and *siniy*—that correspond fairly closely to the English categories of "light blue" (or "baby blue") and "dark blue," respectively. Experiments show that when two color patches fall into different Russian categories—that is, when one is *goluboy* and the other *siniy*—Russian speakers discriminate between them more quickly than when they fall into the same category. English speakers show no such advantage when tested on the same patches. This suggests that Russian speakers judge color differences across the *goluboy/siniy* boundary more keenly than English speakers do, because of Russians' more finely differentiated color terms.

Research of this nature is fascinating and important. But it does not directly undermine the concept of human nature: instead, it seems to alert us to the possibility that culture may be responsible for more of our makeup than we have been inclined to think, and nature less. We are still in the business of playing off nature against culture, even if we think the score-line is a matter for debate. Why, then, have philosophers doubted the propriety of the very notion of human nature?

The Nature of Species

Hull and Ghiselin's skepticism about human nature does not rest on any specific views they have about humans. Their idea is not that human learning, or freedom of will, somehow introduces such protean effervescence into our species that any effort to spell out its nature must be undermined. Instead, their skepticism is based on their general views about biological species of all kinds—cats, cabbages, and coelacanths. They think that the ubiquitous role of variation in the biological world means that no species has a "nature."

A few philosophers have thought that when a biologist asks the question "What makes an organism a member of one species, rather than another?" the answer must appeal to something like genetic makeup.[10] These philosophers have naively assumed that classification in biology works in just the same way as classification in chemistry. Presented with a lump of pure metal, the question "Which chemical element should we assign it to?" is determined by a fact about its internal structure, and more specifically by its atomic number. If its atoms contain 79 protons, then this makes it a lump of gold. If its atoms contain 82 protons, then this makes it a lump of lead. These philosophers have assumed that since membership of a chemical species is determined by a hidden aspect of a sample's internal structure, membership of a biological species must also be determined by a hidden aspect of an organism's internal structure, such as a genetic code.

There has been interminable bickering among eminent biologists over how to understand what sort of a thing a biological species is, but most warring schools of thought agree that species membership is not, in fact, determined by the possession of properties internal to organisms.[11] For example, one very influential account—perhaps familiar from school biology lessons—tells us that a species is a collection of organisms that can potentially breed with each other. If that account is correct, then what makes something a tiger (rather than a dog, say) is not a matter of its having "tiger genes." Instead, it is a matter of its being able to reproduce with other tigers.

There are many other accounts of species membership on the biological market, but most of them concur in rejecting the idea that internal properties are what determine which species an organism belongs to. On some accounts, species are genealogical units of a reasonable size. This view denies that a tiger is an organism with the right sort of DNA: instead, a tiger is an organism with the right parents and grandparents. Other accounts tell us that species are the occupants of environmental niches. This view, too, tells us that a tiger is an organism that makes its living in the right sort of way: it is not an organism with the right genes. All of these accounts tell us that species membership is not a matter of hidden internal constitution. It is a matter of the relations an organism enters into—either with other living organisms, or with past ancestral organisms, or with its environmental niche.

Hull and Ghiselin took the view that anyone who understands this aspect of biological taxonomy will immediately see that species do not have natures. That is because, for Hull and Ghiselin, an organism's "nature" would be the sort of internal property—rather like a chemical element's atomic number—that simultaneously determines which species it is a member of while also explaining the characteristic properties of that species. The nature of gold is fixed by its atomic number: possession of the right number of protons (79) makes something a sample of gold (rather than a sample of lead) while also explaining that sample's electrical conductivity, density, malleability, and so forth. Since there are no biological properties that play both of these roles simultaneously, Hull and Ghiselin argued that no species has a nature.

Hull and Ghiselin augment this basic skepticism of human nature with two further thoughts. First, they point out that it is in the nature of evolutionary processes to make rare traits common, and common traits rare, as new mutations are favored by selection and replace previously dominant traits. Second, they point out that careful research often undermines naive assumptions about the ubiquity of traits within species. We have already seen that psychological research can shake our assumption that people the world over discriminate colors in the same ways, or that they see illusions in the same ways. Charles Darwin's painstaking research on barnacles convinced him that naturalists exaggerate the uniformity of species far too often:

> I am convinced that the most experienced naturalist would
> be surprised at the number of the cases of variability, even in
> important parts of structure, which he could collect on good

authority, as I have collected, during a course of years. . . . Authors sometimes argue in a circle when they state that important organs never vary; for these same authors practically rank that character as important (as some few naturalists have honestly confessed) which does not vary; and, under this point of view, no instance of an important part varying will ever be found: but under any other point of view many instances assuredly can be given.[12]

What does all of this show? If Hull and Ghiselin are right, then we should reject the idea that what makes an organism human is the possession of the right sort of genome. Their work also alerts us to the dangers of complacency in assuming that traits are ubiquitous. Finally, it reminds us that within any given species traits of all kinds can rise and fall over evolutionary time. But none of this rules out a rather more relaxed understanding of "human nature" as a set of features that most humans possess at a time. Indeed, it seems fairly clear that when Pinker and others talk about "human nature," all they have in mind is the collection of traits—especially psychological traits—that evolutionary processes happen to have made very common in our species right now. They can happily acknowledge that these traits might once have been rare, that they do not determine what makes an organism human, that even now not all humans have them, and that they may become rare once again. We have yet to see a good reason to reject this simple conception of what human nature consists in.

Evolution and Variation

The philosopher Edouard Machery has been eloquent in his defense of a modest notion of human nature well suited to the purposes of Pinker and others.[13] In his view, human nature is nothing more than the set of traits made common in our species by evolutionary processes. But even this modest proposal can be challenged on scientific grounds.[14]

First, why think "human nature" must name only those traits that are common? Natural selection sometimes pushes beneficial traits to levels close to 100 percent in a population, but it does not always work in this way. This is a theme stressed by abstract biological theory, and also by observation in the field. On the theoretical side, it has long been understood that if the biological advantage conferred by a trait depends on what other organisms in a population happen to be doing, the result can be a mixed population in which no single type of trait dominates.

This is well illustrated by the evolutionary biologist John Maynard Smith's theoretical treatment of the interactions between "hawks" and "doves." Suppose that organisms are competing with each other for some important resource—it could be food, or mates—and that they act in one of two ways when they run into an opponent. "Hawks" initiate combat, and they do not stop fighting until someone wins. "Doves" back out when faced with aggressive behavior. Now imagine that a population is composed mainly of doves, with just a handful of hawks. Probability dictates that the hawks will usually run into doves, and when they do they will win all their fights by walkover. They will rarely suffer any losses from combat, they will secure significant resources, and, as a consequence, they will prosper and their numbers will increase. This does not mean, however, that doves will be exterminated. For once the population is composed primarily of hawks, then we will find hawks most frequently encountering other hawks, rather than the now-rare doves. The hawks will now get themselves into endless exhausting and dangerous fights, for neither side backs down until someone is injured. It is now the doves who have the advantage, and who consequently increase in numbers, for they avoid combat, and when they encounter other doves they share resources equally between them. The result is a mixture of hawks and doves.

Species that are *polymorphic* (i.e., species that contain a variety of different forms) are not merely predicted by abstract models like the hawk-dove game. Direct observation of nature's diversity also reveals a wonderful array of distinct forms coexisting within single species. The side-blotched lizard (*Uta stansburiana*) is a well-known example from textbooks.[15] The males of the species exist in three distinct forms, each of which possesses characteristic strategic and anatomical adaptations. Males with orange throats tend to be very aggressive. They defend large territories. Other males, with dark blue throats, defend smaller territories and are somewhat less aggressive. Finally, a third kind of male, with a yellow striped throat, does not defend a territory at all. It secures mating opportunities by sneaking into the territories of the others. It appears that no one strategy ever dominates because together they constitute a reptilian game of rock-paper-scissors. The yellow stripes have an advantage over the orange throats, the orange throats have an advantage over the blue throats, and the blue throats have an advantage over the yellow stripes. The three different forms wax and wane, but none ever eliminates the others.

It would be a mistake, then, to think that for every species there must be one single dominant design—a unitary species nature—that natural selection has made widespread. Instead, evolutionary processes can regularly and reliably give rise to species containing a mixed array of forms. It would also be a mistake, as we will soon see, to think that in those circumstances where evolutionary processes do make just one trait highly prevalent within a species, the trait in question is never explained by reference to culture. That point may be obscure when stated in the abstract, but it can be understood by looking at some recent research on human psychology.

Cultural Adaptation

Imitation is a form of learning that involves copying the actions of others. Very few species are able to imitate at all. The primatologist and psychologist Michael Tomasello is skeptical even of the ability of chimpanzees to imitate. He argues that what might appear to be imitative behaviors in chimps are better understood in other ways. If a mother rolls over a log and eats the ants underneath, her infant might notice the presence of ants under the log. The infant might then roll another one over to see if there are more ants there. The infant ends up doing what the mother does, but that is not because she concentrates on her mother's behavior and copies it. She is not imitating.[16] Humans, on the other hand, are excellent imitators. The human ability to imitate is sometimes credited with being a key to the extraordinarily productive features of human culture: by copying what others do, we are able to acquire and then further refine beneficial forms of action. Imitation, on this view, is one of the secrets to humans' spectacular technological progress.

What all this shows us is that imitation is highly developed in humans compared with other species, it seems to be a capacity that more or less all humans have, and it has been exceptionally important in the evolution of our species. For those reasons we are likely to think of imitation as an important feature of human nature. And yet, the psychologist Cecilia Heyes thinks that the capacity to imitate is learned.[17]

One puzzle for theories of imitation is how the growing baby can solve the "correspondence problem." An imitator needs to look at an action in another and then produce a similar one. This might sound easy, but the puzzle of how imitation is achieved is especially acute when one's own bodily movements are hard to observe. If my young son Sam sees me contorting my face in a certain way, how is he able to copy that action? He cannot easily look at his own face to check on whether it is moving in a similar way to mine. What is more, the internal feel of his own face moving does not resemble the look of my own face when it moves in the same way. There is no clear "correspondence" between the look of an action in another and the feel of that same action when one performs it.

Heyes proposes that the links between the perception of an action and the performance of the same type of action can be learned, so long as infants tend to experience performance and perception together. But why should they be experienced together? Heyes gives several suggestions. Sometimes they are associated because babies can look at their own actions. This may be because they inspect their own hands as they are moving them, or it may be through the use of artificial supports such as mirrors. Shared emotional responses—perhaps to a funny-looking situation—can also have the result that when the people a baby is looking at happen to be laughing, the baby will be laughing, too.

Heyes argues that these correlations between the perception of an action and the performance of the same action are enough to ensure that babies are able to associate (a) what an action looks like when someone else performs it and (b) what the same action feels like when the baby performs it. Once these links are established—that is, once the correspondence problem has been solved for simple patterns of action—then more elaborate forms of imitation are possible when complex patterns of these simpler movements are observed together.

Heyes's view has significant evidence in its favor. It helps to explain the facts that chimpanzees can be trained to imitate, that imitation in newborn human infants takes time to emerge, that birds seem able to imitate behaviors that they engage in collectively as flocks, and so forth.[18] This all means that we should take seriously the thought that learning might be responsible for the acquisition of a trait—in this case, the capacity to imitate others—that is not only widely distributed across diverse cultural communities but also of great significance for human interactions and human evolution. Imitation is just the sort of trait that one would presumably want to count as *natural,* and yet its adaptive development appears to rely essentially on *cultural* influence.

The human capacity to imitate appears to be natural, cultural, and a product of evolutionary history all at once. This means that if "human nature" names the traits that have been made common in our species by evolutionary processes, then "human nature" will sometimes pick out traits that have been made common in our species by cultural processes. Cultural processes are a part of evolution. It turns out that the very best conception of "human nature" that we can fashion denies any distinction between what we owe to nature and what we owe to culture.

Untangling Inheritance

Haven't I overlooked something? Is there not a well-established set of scientific techniques for quantifying the respective contributions of nature and the various forms of upbringing? Doesn't the notion of "heritability" tell us the degree to which traits of interest—anything from height to intelligence—are determined by genes? In May 2014, for example, Britain's *Daily Mail* told its readers that "new research shows the ability to recognise faces can be inherited, with 60 per cent of the trait down to genes."[19]

Oddly, the *Mail* chose to run this story under the headline "Find it hard to place a face? It's all in the genes: Inability to recognise people is inherited, study says." The study evidently did not say this inability was "all" in the genes: at best it said just over half of this ability was in the genes, and just under half was elsewhere. But what does this effort to quantify the contribution of genes amount to? Is the idea that, rather in the manner of someone who inherits 60 percent of his wealth from his parents and earns the remaining 40 percent himself, genes contribute 60 percent of an individual's inability to recognize faces, with other factors contributing the rest? This is emphatically *not* what it means to say that the inability to recognize faces is 60 percent heritable. The notion of heritability needs to be handled with extreme care.

Heritability is a technical concept, and it needs to be distinguished from the more familiar notion of inheritance.[20] Very roughly speaking, *heritability* is defined as the degree to which variation across a population in some measurable trait—it could be shoe size, or income—is correlated with variation in the genetic makeup of individuals in that population. As usual, this is easiest to understand with a simple example, removed from the domain of human genetics. Let us begin by thinking about plants.[21]

Imagine we make sure that a field has just the same quality of soil throughout, just the same fertilizer, just the same amount of water and sunlight. And imagine we put genetically different corn seedlings into that field and allow them to grow. Differences in the heights of our mature plants will be fully explained by genetic differences, because their environments are all the same. That means that within this field the heritability of height will be very high. Now suppose we take just one of these plants, make several genetically identical clones of it, and plant them all in a different field where the soil is variable, and where fertilizer and water are sprinkled onto some patches and not others. Again, we let our baby plants grow, and we record their heights. In this field we will find that the heritability of height is very low, because now there are no genetic differences from plant to plant, and variation in height across the field is fully explained by environmental differences.

We can now see more clearly how the technical notion of "heritability" diverges from the informal notion of "something that is inherited." Children typically resemble their parents with respect to the number of fingers they have: parents almost always have ten digits, and their children almost always have ten digits, too. So we might well say that having ten digits is passed from parents to children. But digit number is not a strongly heritable trait. Look across a population, and you are likely to find that most people with fewer than ten digits have suffered accidents with farm machinery, industrial equipment, kitchen knives, and so forth. Some may have been born with lost digits, and so variation in digit number may show a slight correlation with genetic variation, but it is unlikely to be high. It is not a contradiction, then, to say that a trait like digit number is reliably inherited and only weakly heritable.

There are three other important morals to take away from the case of the corn plants if we are to understand heritability in general. First, heritability applies to populations, not to individuals. We can ask what the heritability of height is for the plants in our first field where environmental treatments are constant, or for the plants in the second field where we have planted clones. It makes no sense to ask what the heritability is of height in a single plant. Second, heritability is the sort of thing that can be altered merely by altering individuals' circumstances. If we change the way we look after a field, by making sure that henceforth every plant in that field will enjoy just the same environmental circumstances, then we increase heritability of height in the plants in that field. That is because once environmental differences are eliminated, any remaining differences in height will now be explained by genetic differences. Previously, they were explained by a mixture of genetic and environmental differences. Third, heritability gives us information only about correlations. If we know that height in a population of corn plants is highly heritable, this tells us that differences in height are somehow associated with genetic differences, but taken alone it does not tell us about the processes by which plants with one allocation of genes end up taller than plants with a different allocation.

I can now shed some light on a kerfuffle that broke out in the pages of *The Guardian* in 2013. The story began when *The Guardian* leaked a document written by Dominic Cummings, then special advisor to the British Secretary of State for Education, Michael Gove.[22] In spite of its humdrum title, "Some Thoughts on Education and Political Priorities," Cummings's document contained thoughts on most topics known to man, including complexity theory, weather forecasting, the scientific method, Immanuel Kant, and postmodernism. *The Guardian* focused on a short section of the report in which Cummings had written that genetics had "big potential to inform education policy and improve education."

Cummings argued that "successful pursuit of educational opportunity and 'social mobility' will increase heritability of educational achievement." He was leaning heavily on research by the eminent behavioral geneticist Robert Plomin, and far from distorting Plomin's research (as some hostile columnists tried to suggest), Cummings was offering a reasonable précis of it. Plomin himself went on to argue in a more recent book, co-authored with the psychologist Kathryn Asbury, that "causing an increase in heritability . . . can reasonably be seen as an achievement of which teachers and parents should be proud, rather than a sign of determinism to be mistrusted and feared."[23]

When genetics is linked to social policy, commentators often detect a strong whiff of eugenics coupled with an objectionable form of fatalism that tells us our future is decided inescapably by our genes. Perhaps that was why the UK shadow schools minister, Kevin Brennan, said back in October 2013 that the views of Dominic Cummings "sent a chill down the spine."[24] And yet, when Plomin and Asbury say that an increase in the heritability of educational outcomes is something to be proud of, they are not telling us that genes seal our educational fates. On the contrary, they are alluding to a comforting liberal image of equal opportunity. Their view seems to be that an increase in heritability is a worthy goal because if it is ever attained, that will suggest we have succeeded in evening out differences in educational environments, so that remaining differences in educational achievement can be attributed only to genes.

This might sound sensible, but we should pause. Remember our cornfields. There are lots of different ways of maximizing the heritability of height, because all it takes for heritability to be maximized is for all plants to be exposed to the same environment. A farmer would hardly be proud to know that a population of her corn plants shows very high heritability if that is because every one of them has too little water, poor soil, and no fertilizer. What is more, when we think about how we might change this impoverished environment, there is no guarantee that interventions that are good for one corn plant will be good for all. Idiosyncratic differences might mean that one corn plant will flourish best in horse manure while another will flourish best in cattle dung. If what we want is for all of our plants to thrive to their full potential, we may well have to treat them differently. That means exposing them to different environments, and thereby reducing heritability. So it is hard to see why high heritability is something to aim at for educational outcomes. To say that heritability of educational outcomes has been maximized in a population does not mean that children's potential has been maximized.

Plomin is aware of all this, and this awareness makes his remarks about reacting with pride to an increase in heritability puzzling. He himself stressed in an interview with *The Guardian* that "children differ in how they learn," and his book with Asbury makes clear that high heritability could potentially be a result of children all being exposed to similar poor teaching methods.[25] For that reason, Plomin is also amply alert to the distortions involved in thinking that heritability quantifies the extent to which nature, rather than culture or society, is responsible for some inequality in achievement. High heritability of educational outcomes is compatible with everyone failing to reach their full potential, because no one is properly taught.

If we want our schools to bring out the best in children, we need detailed information about the differing methods and mechanisms by which children can be motivated, and by which they acquire valuable knowledge and skills. Heritability studies give us information about correlations between genotypes and educational success. Perhaps one day we might be able to transform this correlational knowledge into detailed insight about the processes by which learning works. That knowledge might eventually permit more effective interventions to cater to the learning needs of all. But we are a very long way from that stage of scientific maturity.

The Natural Order

Psychological research suggests that many humans are intuitively attracted to a particularly troublesome conception of species' natures. It seems that young children tend to think that each living kind has an internal nature of sorts that, when it is functioning properly, produces the typical visible features we associate with the species in question.[26] In other words, they think of cats as all possessing some hidden inner property that is outwardly manifested in typically "catty" behaviors like hunting mice and purring on laps. These internal natures can consistently misfire, with the result that their proper effect does not materialize. All true cats possess the underlying cat nature, but any number of those true cats may fail to hunt, or to purr.

There is further evidence that people are prone not only to think of biological species as having hidden essences but also to think of sexes and races in these ways.[27] This image of an internal essence, shared by all members of a sex or all members of a racial group, which may or may not be manifested in outward behavior, is arguably what underlies many harmful racist or sexist stereotypes. It is what allowed Darwin to give generic characterizations of "the negro" and "the Australian." It is what enabled him to quote with approval William Greg's description of "the careless, squalid, unaspiring Irishman." If all "Irishmen," or all "negroes," share a common nature, then it makes sense to offer a unitary description of that nature. If those internal natures can misfire under unsuitable circumstances, then their existence cannot be disproven by pointing to Irishmen who happen to be careful, affluent, or full of ambition. This notion of an internal, essential nature does harm, in part because it is so resistant to evidence.[28]

Leon Kass's remarks on the wrongs of human cloning demonstrate the discomfort induced by trying to discuss ethics using the language of human, or mammalian, nature. Some of his writings make the compelling point that the mere fact that some trait is bestowed on us by apparently natural processes tells us nothing at all about how we should evaluate that trait. Evolution may have equipped humans with some traits we need to cultivate, others we would be better off without.[29] What, then, could Kass be aiming to achieve by telling us that "sexual reproduction . . . is established . . . by nature; it is the natural way of all mammalian reproduction"?[30] This cannot amount to a case against asexual forms of cloning, unless supplemented by some further argument that explains why sexual reproduction is to be celebrated and why asexual reproduction should be frowned upon.

To his credit, this is just what Kass attempts to do in his argument against the moral permissibility of human cloning. He tells us that a colleague once asked what his position would have been if the standard means of human reproduction had been asexual and scientists had invented a novel technique whereby sexual reproduction might become possible. Would he have opposed efforts to alter human nature by making us sexual reproducers? Kass hints that he would not have stood in the way of this innovation, for his view is that sexual reproduction is morally admirable. Asexual creatures face a cruel world of alienation. For sexual creatures, things are altogether more heartening: "For a sexual being, the world is no longer an indifferent and largely homogeneous otherness. . . . It also contains some very special and related and complimentary beings, of the same kind but of the opposite sex, toward whom one reaches out with special interest and intensity."[31] It is hard to know how seriously to take this defense of sexuality. Plants, of course, are sexual beings. Some plants also regularly reproduce asexually, through the production of runners. Are apple trees less prone than strawberries to *Weltschmerz*?

Kass casts further doubt on the respectability of asexuality by telling us that we "find asexual reproduction only in the lowest forms of life: bacteria, algae, fungi, some lower invertebrates."[32] Kass's list is incomplete. I have already alluded to the many plant species that can reproduce asexually. Parthenogenesis—a form of asexual reproduction whereby female eggs develop without fertilization from males—is regularly observed in reptiles.

Setting aside these qualms about the depravity of asexuality, the prevalence of asexual reproduction in bacteria does not carry any moral weight when we try to think about the wisdom of allowing cloning in humans. Kass rightly wishes to cultivate the desires of humans to reach out to each other with "special interest and intensity." Some humans do this without having children at all; others reach out to members of the same sex, and they raise adopted children in the process. Sexual reproduction is not necessary for the feelings Kass admires, neither is it sufficient. Plenty of reproduction occurs accidentally, some occurs irresponsibly. So let us imagine two women who love each other, and who wish to start a family. They both want an intimate biological role in the process, and they achieve this by placing a cloned embryo from one partner into the womb of the other. If Kass is concerned to preserve a world in which rich relationships flourish between other-regarding humans, more argument is needed to show that asexual reproduction like this will undermine it. Appeals to the nature of humans, or the nature of mammals, will not do the trick.

The Dangers of "Human Nature"

We might have thought that "human nature" was a problem-free concept—one that innocently picks out the universal, evolved features of our species. But we have now seen how much trouble "human nature" can get us into. It is a mistake to think that all evolved traits are universal—that was the lesson of our exploration of evolutionary polymorphism. It is also a mistake to think that universal traits cannot be learned—that was the lesson of Heyes's work on imitation. We have observed the confusion caused by "human nature" when it finds its way into ethical discussion, and the ways that thinking in terms of a group's "nature" can reinforce racial and gender stereotypes. The sciences do not need a concept of "human nature" if they are to understand the processes that have introduced patterns of similarity and difference into the psychological makeup of our species. If they do not need "human nature," and if it repeatedly causes problems, we would be better off avoiding it altogether.

Further Reading

The best overview of work on the concept of human nature is:
Stephen Downes and Edouard Machery, eds., *Arguing About Human Nature: Contemporary Debates* (London: Routledge, 2013).

For a vigorous effort to show the pervasive role of culture in human development, see:
Jesse Prinz, *Beyond Human Nature: How Culture and Experience Shape Our Lives* (London: Allen Lane, 2012).

For an equally vigorous defense of human nature, see:

Steven Pinker, *The Blank Slate: The Modern Denial of Human Nature* (London: Penguin, 2002).

For an introduction to the role of culture in human evolution, see:

Peter Richerson and Robert Boyd, *Not by Genes Alone: How Culture Transformed Human Evolution* (Chicago: University of Chicago Press, 2005).

Chapter Eight

Freedom Dissolves?

The Myth of Choice

Naive intuition need not line up with how things are. The Earth doesn't look, to observers standing on its surface, much like a squashed sphere, but it is one. Whales don't look much like mammals—at least not on the outsides—but they are. Some of the most arresting scientific discoveries—and, for that matter, some of the most striking pieces of historical and literary research—show how far the true workings of the universe and its inhabitants depart from our untutored expectations. Even so, one might wonder if sometimes scientists go too far in announcing the debunking of widely held myths.

It seems that we regularly face choices about how best to act, and that conscious reflection helps to determine which course of action we end up taking. It seems, in other words, as though we often have a kind of freedom regarding what we do. When I last bought a car, I spent quite a bit of time over my choice. I got advice from friends, looked at websites, considered my budget, consulted with my spouse, took a test-drive with my daughter.

The whole process demanded mental effort, and the car I came away with was the one I decided (for better or worse) was right for us. It would be very surprising, then, to learn that my strong impression of conscious deliberation making a *difference* to this outcome was, in fact, a mistake, and that none of my patient reasoning had any impact on which car ended up outside my house. And yet this is just the sort of thing that scientists have been lining up to tell us in recent years.

To take just one example, in a widely cited paper in *Nature Neuroscience* from 2008, John-Dylan Haynes and colleagues provided evidence which, they said, indicated that our "subjective experience of freedom is no more than an illusion."[1] Many others have endorsed this verdict. Or rather, it seems they have endorsed it. For in the domain of free will, just what is being asserted and just what is being denied are slippery matters. The atheist and science writer Sam Harris draws on scientific work to argue that "free will is an illusion. . . . We do not have the freedom we think we have."[2] So do we not have free will at all, or is it just that freedom isn't quite what most of us take it to be? The neuroscientist Patrick Haggard told readers of the UK's *Daily Telegraph* that "we certainly don't have free will. . . . Not in the sense we think."[3] But what is that sense, and do we have free will in some other more sensible sense?

Michael Gazzaniga, another neuroscientist of exceptional distinction, informs us that "neuroscience reveals that the concept of free will is without meaning, just as John Locke suggested in the 17th century. . . . It's time to get over the idea of free will and move on."[4] If the concept of free will is literally without meaning, then it makes no more sense to deny we have it than to assert we have it. So, has scientific research really imperiled the notion that our conscious deliberations often make a difference to what we end up doing?

These recent assaults on free will by prominent scientists draw on two quite distinct lines of attack, and it is worth evaluating them separately. First, there is a very general form of skepticism that does not draw on any new scientific evidence, but which is instead based on a worry centuries old. We can call this the *causal nexus* argument. If human bodies, and human minds, are parts of the causal order of the universe, then it seems to many that we can be nothing more than conduits for chains of influence that originate in times and places that are quite external and alien to us. Humans might cause crimes, wars, and greenhouse gas emissions, but humans are never authors of these actions, just as an avalanche is not an author of the devastation it brings. The avalanche itself is merely a consequence of the earlier snowfalls and the triggering conditions that set it off, and human actions are merely consequences of whichever past social, neural, and genetic conditions happened to produce them.

It would seem that the only way to restore free will to humanity—the only way, that is, to restore a justified conviction that we are not just passive systems of pipework through which energy flows, but that we are in charge of how things turn out—is through a heroic denial that humans are part of nature. Gazzaniga's skepticism of free will is largely based on considerations of this sort. He thinks that a defender of free will must find a way to argue that humans are free from the causal-mechanical order of things; that we are somehow exempt from the usual pushes and pulls that characterize how material objects behave. It is not surprising that he does not wish to swallow this pill. After all, the successes of neuroscience are founded on the assumption that our actions depend on how our brains are configured, and on the assumption that the states of our brains, like the states of other natural systems, are causally influenced by prior internal and external states of the world.

The second line of scientific attack on free will is different. It is both more specific in its focus and more recent in its genesis. Instead of drawing on general conceptual argumentation, it draws on particular experimental results. In that sense, it makes more constructive use of novel scientific data than the causal nexus argument. We can call it *the argument from tardiness,* for reasons that will soon become clear.

John-Dylan Haynes and his group, whom I mentioned briefly at the beginning of this chapter, reported in their 2008 paper that they were able to predict, using information from a brain scanner, which action an individual would choose up to ten seconds *before* the individual made a conscious choice to act.[5] Their experiment updated and extended striking and seminal work by Benjamin Libet, whose own research suggests that decisions to act one way rather than another "bubble up" (as Libet puts it) from unconscious brain processes, with the result that conscious awareness of a decision to act comes well *after* the act is initiated.[6]

Libet's and Haynes's results are often thought to show the illusory nature of free will. If something in our brains initiates a specific course of action, and if it is only later that we have an impression of consciously choosing that same course of action, then it appears to many that our conscious decisions cannot truly influence what we end up doing. Rather like the person who trips by accident on the street and then gives the absurd impression of having intended to give a slapstick performance all along, our conscious intentions are impotent retrospective endorsements of pathways we are already irretrievably committed to.

The Causal Nexus Argument

The causal nexus argument is familiar to philosophy students the world over. Its appeal derives from a simple dilemma. It seems that "freedom of the will" involves a form of control, such that we can steer the course of our actions one way or another, depending on our inclination. Is there any reason to think we have such a capacity for control? Either our actions are causally determined by a constellation of prior events inside and outside our brains or they are not. If they are *not* determined by a constellation of prior events, then this seems like bad news for control. After all, we do not want our actions to take the form of spontaneous ejaculations, of the sort that might take us entirely by surprise.

Control seems like the sort of notion that is best understood in terms of the causal influence of our deliberations and so forth over what we end up doing. But if our actions *are* determined by a constellation of prior events, then it seems there is no room for us to intervene in such a way that things go one way rather than another. Given that prior constellation, our later course of action is already decided. Jerry Coyne, an accomplished evolutionary biologist, relies on this argument in his own skeptical treatment of freedom:[7]

> If you could rerun the tape of your life up to the moment you make a choice, with every aspect of the universe configured identically, free will means that your choice could have been different. . . . Although we can't really rerun that tape, this sort of free will is ruled out, simply and decisively, by the laws of physics.

In spite of my impression to the contrary, when I bought my Ford, I really wasn't free to buy a VW. I wasn't even free to buy the same model with different trim: reset the universe to my birth in 1974, press play again, and that very same Ford will always appear in front of my house in February 2011.

A World of Chance

It would be understandable if, at this point, one were to complain that arguments against free will rely on outdated science.
Quantum physics tells us that the universe is a chancy place.
Physicists and philosophers say it is *indeterministic*. In a *deterministic* universe, the laws of nature have the result that a
complete snapshot of how everything is at one time fixes every
future state of that universe for the rest of time. In a deterministic universe, the configuration of events just before the big bang
is compatible with just one subsequent evolutionary pathway
for that universe. In an indeterministic universe, things are
more relaxed. Quantum physics says that if we bombard an
unstable radioactive nucleus with energy, then while we might
make it much more likely that the nucleus will decay in the near
future by emitting an alpha particle (i.e., a particle that contains
two protons and two neutrons), our actions do not guarantee
that alpha emission will occur at any specific time in the future,
or even that it will occur at all. In an indeterministic universe, a
complete snapshot of how everything is at one time is compatible with that universe evolving in many different ways for the
rest of time. If our own universe is indeed indeterministic in
this way, then, on the face of things, we can rerun Coyne's tape
many times, and we will find different cars parked on my drive.

The problem with this response to the free will problem returns us again to the issue of control. Let us agree that our universe is indeterministic. For the sake of argument, we can even
agree with the more contentious claim that this indeterminism
is not only manifest in the quantum domain but also "perco-
lates up" (as the phrase goes) to the level of everyday observable
events. The question is what the significance of this kind of indeterminism might be for freedom.

What we want from free will, it seems, is to secure the claim that we are in control of things. This is not what indeterminism gives us. Instead, indeterminism suggests that, just as an excited atomic nucleus may or may not decay in a five-minute interval, so a resolved car buyer may or may not purchase a Ford in a five-minute interval. But indeterminism does not tell us that the atom is in charge of when it decays, and it does not tell us that a person is in charge of whether she buys a Ford. Indeterminism gives us some reason to think that several alternative futures may be equally open to a deliberating individual, just as several alternative futures are equally open to the excited atom.[8] But it is chance, not control, that dictates which of these futures materializes.

If it is control over the future that we want to secure, it is difficult to see how an appeal to indeterminism can help us. For that reason, many commentators on free will, even as they accept the reality of indeterminism, have taken the view that it offers little by way of hope for freedom. Instead, the question at stake when we think about the reality of freedom is whether the causation of later events by earlier ones poses a problem for the idea that our conscious deliberation makes a difference to how things turn out. Whether causation is of the indeterministic variety, which merely increases the chances of these later events, or whether it is deterministic, which guarantees them, is irrelevant.

Natural Freedom

In recent years philosophers have tended to argue that the causal nexus presents no genuine problem for freedom. Of course, the causal nexus does present a problem if "freedom," and "free will" are defined in ways that make them inherently spooky, supernatural, or otherwise incompatible with an image of persons as organisms thoroughly entwined in causal inter- action with their environments. If, for example, the assertion that we have free will is understood to mean that we have wise little people inside our heads who have the power to sponta- neously formulate plans of action that are cut loose from any prior causal influence, and whose decisions determine the sub- sequent behaviors of our bodies, then of course free will is an illusion. But, as Daniel Dennett wisely advises us to ask, what reason do we have for thinking that this would be the only sort of free will "worth wanting"?[9]

A different way of understanding what freedom involves casts the free agent as someone with a distinctive set of ca- pacities. A free agent is able to respond in appropriate and flexible ways to her surroundings, without physical imped- iment or restraint. A free agent is able to contemplate and weigh rational considerations, initiating a course of action if her deliberations suggest it is suitable. I am free, then, if I am the sort of organism that is sophisticated enough to process information about a car's price and its fuel economy; to form aesthetic preferences and ascertain whether they have been met; to check on whether upholstery can withstand violent treatment at the hands of young children; and then to execute a plan for purchase without intimidation, coercion, and so forth. The sciences do not deny that there are such organisms: instead, many sciences are actively engaged in determining the extent to which different species—primates, birds, hu- mans—are capable of different forms of plastic response, the extent to which they process information in suitably sensitive ways, and the plausible rationales for the emergence of such sophisticated capacities.

Everything You Ever Wanted to Know About *Sphex*

There is a world of difference between an organism that responds in rigid, routine ways to environmental stimuli and an organism that is, instead, exquisitely sensitive to the details of its predicament. Research undertaken on the digger wasp *Sphex*, described by Dean Wooldridge in the 1960s but brought to the attention of researchers on free will in a series of Dennett's philosophical works, is often used to illustrate just this point.

Before laying her eggs, *Sphex* builds a burrow and finds a cricket. She does not kill it. Instead she paralyzes the cricket with her sting, drags it into her burrow, lays her eggs next to it, and flies away for good. Once the eggs have hatched, the grubs have a supply of fresh cricket on which to feed. This all sounds like sensible behavior on the part of *Sphex*. But Wooldridge went on to describe *Sphex*'s failings:[10]

> The wasp's routine is to bring the paralyzed cricket to the burrow, leave it on the threshold, go inside to see that all is well, emerge, and then drag the cricket in. If, while the wasp is inside making her preliminary inspection, the cricket is moved a few inches away, the wasp, on emerging from the burrow, will bring the cricket back to the threshold, but not inside, and will then repeat the preparatory procedure of entering the burrow to see that everything is alright. If again the cricket is removed a few inches while the wasp is inside, once again the wasp will move the cricket up to the threshold and re-enter the burrow for a final check. The wasp never thinks of pulling the cricket straight in. On one occasion this procedure was repeated forty times, always with the same result.

Poor *Sphex* is rigid: a simple manipulation of the environment shows that she merely executes a fixed pattern of behavior, never considering that because she has already checked her burrow, she should simply get on with provisioning her larder. We humans, so the story goes, are flexible. When we act we attend closely to our own past actions, to the actions of others, to the nature of our surroundings, and we chart a sensible course that is sensitive to all of this. It is hard to see what more we could want of freedom, and freedom, in this sense, is both a respectable object of scientific investigation and an evolutionary achievement restricted to only a few species at best. In understanding that *Sphex* is trapped, we also understand that we are free. Freedom, says Daniel Dennett, evolves.

There is an irony to the *Sphex* story, beautifully exposed in recent historical work by Fred Keijzer.[11] *Sphex*'s behavior turns out to be more variable, more sensitive, and more sensible than folklore about free will suggests. Wooldridge, Dennett's source for his descriptions of *Sphex*'s behavior, was an engineer in the aerospace industry. He never did research on insects himself. It seems that Wooldridge got the *Sphex* story from a 1938 edition of *The Science of Life,* a popular summary of biological knowledge by H. G. Wells, Julian Huxley, and G. P. Wells.[12] They, in turn, were reporting on original research first described back in 1879 by a Frenchman, Jean-Henri Fabre, finally published in English in 1915 as *The Hunting Wasps.*[13]

Fabre did indeed report moving a cricket, left at the entrance to *Sphex*'s nest, by a few inches, and watching her drag it back to the entrance before disappearing back into the nest. He repeated his trick forty times, "and her tactics never varied." But Fabre was uncomfortable:[14]

What I asked myself was this: "does the insect obey a fatal tendency, which no circumstance can ever modify? Are its actions all performed by rule; and has it no power of acquiring the least experience on its own account?" Good fortune brought me into the presence of another colony of Sphex-wasps, in a district at some distance from the first. I recommenced my attempts. After two or three experiments with results similar to those which I had so often obtained, the Sphex got astride of the cricket, seized him with her mandibles by the antennae and at once dragged him into the burrow. At the other holes, her neighbours likewise, one sooner, another later, discovered my treachery and entered the dwelling with the game, instead of persisting in abandoning it on the threshold to seize it afterward.

Fabre himself immediately found variation in *Sphex*'s behavior from colony to colony: these wasps were not all doomed to repeat the same fixed cycle of behavior.

A much more recent study of the great golden digger wasp, *Sphex ichneumoneus,* was published by Jane Brockmann in 1985.[15] She, too, found that individual wasps varied in terms of how likely they were to quickly break the cycle and bring a katydid (a close relative of the cricket) straight back into the nest after she had moved it. She also suggested that the wasp's repositioning of a moved katydid was in fact an advantage to the wasp: if the katydid is not facing headfirst toward the nest, then when the wasp tries to drag it in it is likely to get stuck. The wasp needs to reenter the nest and come back out again headfirst, because she needs to drag the katydid into the nest by its antennae. It shouldn't be surprising, then, that each time an experimenter repositions the katydid, the wasp first has to place it at the entrance to the nest, before going back into the nest to turn around and stick her head out.

Brockmann argues that *Sphex*'s behavior is variable, versatile, adaptable, and generally more sensible than philosophical myth would have us believe. Keijzer remarks pithily that while it remains uncertain to what degree *Sphex* is doomed to repeat an endless cycle of behavior, it seems much clearer that we philosophers have been guilty of endlessly repeating a simplified version of the *Sphex* story, without paying due attention to its complexities and its genesis.[16]

Compatibilism

The realities of the *Sphex* case show that we should not be too quick in assuming that insects are rigid whereas we are plastic. Sometimes humans settle too quickly into behavioral ruts: they churn out the same stories about what wasps do, regardless of how appropriate these stories are. Sometimes wasps show sensible strategies in the face of the complex logistics of getting a cricket into a nest. But this doesn't undermine the most important aspect of Dennett's discussion of free will; it reinforces it. The question of how flexible animal and human behavior might be is something we care about, and it is something open to scientific investigation. A demonstration that we are unfree would require patient work, showing how our choices are insensitive to the details of our circumstances.

A large amount of experimental work in comparative cognition, for example, asks about the degree to which species of ape are able to attend not only to other apes' pieces of behavior but to their mental states as well.[17] These debates are lively, and they have forced researchers to be extremely creative when it comes to designing experiments. On Dennett's view, scientific investigation might show that I was condemned to purchase a Ford if it could show that at the time I made my purchase I was subject to a kind of irrational love affair with the Ford Motor Company, such that no amount of adverse information would have swung my decision another way, or if it could show that I harbored a secret terror of all other manufacturers. Indeed, psychological research frequently does show that humans are prone to neglect certain kinds of information that are relevant to their decisions, and that they overestimate the subtlety of their own thought processes.[18]

This sort of work draws our attention to the ways in which we are more rigid, less flexible, than we might think at first. But we do not learn that we lack freedom simply by learning that our actions are caused: that is because the capacity to respond flexibly and appropriately needs to be instantiated in complex causal mechanisms. Instead, our freedom is undermined only to the extent that we discover that our actions are caused in particular, rigid, ways. Our ability to respond appropriately to each other and to our environments is undermined to the extent that we are insensitive to changes in circumstance or changes in evidence. Even then, freedom is undermined only by degrees. Freedom and causation, in this view, are compatible.

Plenty of commentators have been unimpressed by this "compatibilist" response to the problem of free will. Some critics tell us that while perverse philosophers might think that a free individual is the sort of organism whose complex causal capacities give it a certain kind of unfettered sensitivity to its surroundings, this is not what most people have in mind when they are trying to decide whether they are free. What most people consider to be freedom, we are told, is instead tied up with the far more spooky, scientifically intolerable idea that human action is independent of prior causal influence. So Sam Harris, for example, dismisses compatibilism on the grounds that, in his view, it solves a problem that no one (academic philosophers excepted) cares about, while leaving unsolved the pressing problem that bothers the general public.

This helps to explain why, in recent years, discussion over free will has often turned to the psychological question of what people on the street mean when they say they have free will. There are reasons to be wary of this twist in the debate. The problem of freedom of the will is, and has always been, somewhat rarefied. This should make us skeptical of studies that ask what "most people" understand by freedom. Perhaps we can persuade people that they have opinions about these matters, but it is unlikely that many people have thought about them enough to come to anything like a considered view. It is always the case that survey data must be taken with a pinch of salt, especially when people are asked to give simple responses to complex questions. If you stop people on the street and ask how much money they would be willing to pay to avoid a year with bowel cancer, they will often give you an answer. It does not follow that they have ever thought much up to that point about what life would be like with that disease, or how to translate this into a monetary equivalent.[19]

Returning to the domain of free will, several philosophers have claimed that humans are "natural incompatibilists." Humans, they say, are naturally inclined to think that if our actions are determined by prior causes, then those actions cannot be free.[20] The further implication is that it takes considerable philosophical argumentation—perhaps contrived, almost certainly unconvincing—to talk people out of this common-sense view. But "incompatibilism" names a rather technical set of views. It is the denial of the compatibility of two claims: first, that we act freely; second, that the universe's laws specify a unique pathway for its evolution over time. It is hard to imagine anyone coming to an opinion about that doctrine "by nature."

Let us set these caveats to one side. The question of what most people understand by "free will" is an empirical one: we cannot assume we know the answer without carrying out some investigations on real people. Eddy Nahmias and colleagues conducted some surveys that tried to answer this question, and the conclusion they arrived at looks like bad news for those who think that compatibilism is the sort of misshapen, artificial position that only someone tainted with philosophy could ever adopt.[21]

For example, Nahmias and his collaborators asked people to imagine that a supercomputer might be able to predict the future so accurately that it could tell us, twenty years before a person was born (they call him Jeremy), the precise moment at which Jeremy would rob a bank. If a computer can make this prediction so far ahead of time, it seems that Jeremy must live in a deterministic universe. They then asked these people, "Do you think that, when Jeremy robs the bank, he acts of his own free will?" A full 76 percent of participants said yes, he robs the bank of his own free will. A smaller majority—67 percent—also judged that Jeremy could have chosen *not* to rob the bank, in spite of the predictability of his action. So it seems only a minority of them took the view that determinism would stand in the way of freedom.

This piece of work can certainly be challenged. Are we sure, for example, that Nahmias's respondents really believed Jeremy's universe to be deterministic? Did they believe, in other words, that the supercomputer was making its predictions by calculating the consequences of deterministic laws twenty years ahead of time? Or did they simply think that the supercomputer had the magical ability to peer directly into the future of an indeterministic universe?[22] Nahmias's work is far from conclusive, but it does make a suggestive contribution, once the ground of debate has shifted to the claim that what most people mean when they discuss "free will" is not the sort of freedom the compatibilist might secure. What this survey work does not do, of course, is to tell us about whether these compatibilitist views are sensible. That is the question we must now turn to.

Could You Have Done Otherwise?

Can any sense be made of the idea that, in spite of Jeremy's actions being consequences of deterministic laws, he could nonetheless have done otherwise? Plenty of commentators have thought this to be a piece of compatibilist sophistry. Determinism tells us that the future is an inevitable consequence of the past. If what we do is inevitable, then, it seems, we couldn't have done otherwise. If we could not have done other than we did, we are not free. The philosopher must resort to what Sam Harris disdainfully refers to as "theology"—on the grounds that it is a polished and professional defense of the contradictory and absurd—to argue that determinism does not stand in the way of freedom.

Compatibilist theology begins with common sense. I bought a Ford. I would have liked a VW, but I didn't buy one because it was much more expensive than the Ford. Could I have done otherwise? Could I have bought a VW? I had just about enough money to buy a VW: what stood in the way of my buying one was that I felt it would be wiser to spend that money on other things. Had I been less concerned about paying off my mortgage and covering nursery fees, and more concerned about German engineering, then I would have bought a VW. In other words, I could indeed have bought a VW had my priorities been different. That claim is not remotely threatened by determinism—no more than determinism threatens the claim that an arrow could have traveled a little further if it had been shot with greater force.

"But," says the skeptic, "this is changing the subject. A free person isn't someone who could have done otherwise had things been a bit different. A free person is someone who could have done otherwise even if things had been exactly the same. Determinism tells us that, for any fully specified state of the universe at a time, there is a unique future course of evolution for that universe. Determinism is incompatible with free will because it is incompatible with the claim that one could have done otherwise." We have already seen that Jerry Coyne appeals to this sort of reasoning in his attack on free will:[23]

> I construe free will the way I think most people do: At the moment when you have to decide among alternatives, you have free will if you could have chosen otherwise. To put it more technically, if you could rerun the tape of your life up to the moment you make a choice, with every aspect of the universe configured identically, free will means that your choice could have been different.

I have suggested that we should pause before accepting any claim that "my" construal of free will is probably the same as "most people's." Coyne is suggesting that for most people, free will is the sort of thing that is inevitably ruled out by determinism, and we have seen that there is some experimental evidence against this claim about how most people see things. Suppose, though, that Coyne is in alignment with the people: Is their view of the matter a sensible one?

Determinism leaves untouched the idea that humans are impressive organisms that can consult evidence, weigh reasons, and formulate a plan for action. These capacities are sensitive to the fine details of local circumstance, which is just another way of saying that had their inputs been different, the results of deliberation would have been different, too. This allows us to appreciate an important sense in which determinism leaves our actions unconstrained. A constrained process will tend to the same endpoint, regardless of how it starts out. A ball bearing will roll to the bottom of a teacup regardless of where on the rim its release point is located. In that sense, it is inevitable that the ball bearing rolls to the bottom. Our actions are not constrained in this sense: how things end up depends in very fine-grained ways on how they begin. Moreover, these dependencies are often rational: what we end up doing depends on where the evidence points. Had the evidence pointed elsewhere, we would have acted differently. All of these things involve forms of freedom, all are compatible with determinism. It is not clear why we should require anything more.

The Argument from Tardiness

We have seen that some recent efforts on the part of scientists to undermine free will turn on very general—and very old—worries about the relationship between freedom and causation. A more novel set of considerations against free will comes from a tradition of experiment in the neurosciences, which seems to show that conscious deliberation has no significant effect on what we end up doing. A falling barometer may precede rain, but a falling barometer does not make the rain come. Instead, the falling barometer and the rain are both effects of a common cause—namely, a drop in air pressure. Similarly, our conscious decisions may precede our actions, but they do not make us act one way rather than another. Instead, our conscious decisions and our actions are both effects of a shared set of earlier causes in our brains. Or so the story goes.

These claims are based on experiments whose goal is to investigate the timing of what are called "spontaneous" decisions—that is, decisions to do something for no particular reason other than that one fancies it. In Benjamin Libet's classic early experiment in this genre, his subjects were asked to flex their wrists whenever they felt like it.[24] Of course, most of our decisions are not spontaneous in this sense. Typically when I flex my wrist it is not because I have an urge, coming from nowhere in particular, telling me that wrist flexing is the thing to do. Instead, I flex my wrist because I need to knock on a door, tap someone on the shoulder, and so forth. These decisions to move my wrist are made with good reason. They are prompted by a closed door, or a turned back.

Psychologists have discovered that before spontaneous voluntary movements of the wrist-flexing sort occur, there is an increase in neuronal activity known as the "readiness potential," or RP for short. It is possible to measure the onset of the RP using something called an electroencephalogram (EEG). In his experiment Libet asked people to flex their wrists at whatever time they felt the urge, ignoring any external stimuli. He then recorded the timing of three events. First, he needed to record the time at which his subjects felt the conscious urge to flex their wrist. He did this by asking his subjects to look at a clock with a rotating dial and to recall the position of the dial at the moment they felt the urge to flex. Second, he used an EEG to record the time of onset of the RP, which he understood to be the time at which flexing was initiated by the brain. Third, he measured the time of flexing itself.

What Libet found is, perhaps, surprising. He found that the onset of the RP comes about 550 milliseconds (i.e., about half a second) before action. But he also found that the RP comes about 350 milliseconds before subjects reported feeling the urge to flex their wrists. In other words, the RP comes first, then the conscious urge to flex, then the movement of the wrist. This means that someone monitoring the onset of the RP could predict that a person will flex her wrist before that person even feels the urge to do so. And that, in turn, has led many commentators to suggest that the person's conscious decision comes too late for it to be a cause of her wrist flexing.

Libet's experiments are fascinating, and they have generated a large literature offering a variety of different interpretations.[25] Let us grant the basic chronology Libet suggests: neuronal activity increases, subjects feel the urge to flex, and then subjects flex. Let us also grant that this increase in neuronal activity is a reasonably good predictor of when flexing occurs. Does it follow that the urge to flex is not a cause of flexing? It seems to me that it does not.

Consider this ordering of events: first, the starter's gun goes "BANG!"; second, Usain Bolt feels an urge to accelerate out of the blocks; third, Bolt runs. Once the gun has been fired we can predict with great reliability that Bolt will accelerate, even though there will be a very small window of time before Bolt himself forms any urge to do so. The firing of the starter's gun comes before Bolt's urge to accelerate, but that does not mean that Bolt's urge is impotent. He runs because he feels the urge. And he feels the urge because he hears the gun. The urge comes a little after the gun, but that is just because it takes time for the sound to reach him, and it takes time for him to react.

What does the Bolt case have to do with the Libet case? Bolt's acceleration is not a "spontaneous" action. Bolt doesn't accelerate just because he feels like it; he accelerates because he hears the gun. "Spontaneous" actions, on the other hand, are supposed to issue entirely from the inside: they are not meant to be prompted by external stimuli. Now, if you are asked to flex your wrist without attending to external stimuli, but merely when you feel the urge, it is quite possible that you may never feel the urge at all, and your wrist will remain annoyingly straight.[26] But experimental subjects are meant to flex their wrists at some time or another while the experiment is under way—if they do not, then Libet cannot get any data.

How can subjects make sure they obey the experimenter's instructions if they are worried that they aren't the sorts of people who ever have spontaneous urges to flex their wrists? Here is one solution: let some rumbling of your brain act as a sort of "BANG!" signal, and resolve that you will flex your wrist when you hear the "BANG!" If this is what happens in Libet's experiments, then perhaps the RP is a form of internal "BANG!" It prompts the urge to flex one's wrist, and that urge in turn causes the wrist to flex. Just as we are not surprised that the starter's gun precedes Bolt's urge to accelerate, so we should not be surprised that the RP precedes the urge to flex.

This sort of speculation sounds suspiciously like armchair neuroscience. Can it be backed by evidence? First, a recent piece of detailed neuroscientific work has suggested that the nature of the RP may have been misunderstood. Scientists have tended to think of the RP as a neural indicator of something akin to a plan to move: the RP indicates an unconscious determination to flex one's wrist, for example. New experimental work has made trouble for this interpretation. A team in New Zealand asked subjects to wait until they heard an audio tone before deciding whether or not to tap a key. The team reasoned that if, as Libet suggests, the RP is an indicator of upcoming action, then they should detect an RP when people do decide to tap, but not when they don't. Instead, they detected an RP regardless of what their subjects ended up choosing.[27] So the RP does not seem to be an unconscious resolution to move.

Second, an even more recent study by Aaron Schurger and his collaborators has suggested a detailed account for how the RP is generated, which is consistent with the idea that it is a kind of internal "BANG!"[28] Normally, we make decisions based on the accumulation of evidence. Once there is a decent amount of evidence pointing in favor of a course of action, we proceed. As we have seen, Libet's subjects were asked to take on a peculiar sort of problem: to flex their wrists when they felt the urge, in spite of the fact that there was no relevant source of evidence that might tell them that now is a good time to flex. Most of our decisions are not like this. Bolt goes when he hears the gun, not whenever he fancies. Whimsical decisions, too, are typically prompted by suitable reasons and suitable opportunities. Even when I "just fancy" an ice cream, the fact that I buy one is prompted by the warm weather, some time away from work, and the proximity of a stall that sells them. In short, the sort of task people are given in Libet's experiments is highly unusual.

Schurger and colleagues suggest that when presented with this artificial task, we simply let "physiological noise" determine whether to act. In the absence of any sensible cue that might prompt the flexing of a wrist, we allow the random background burbling of our neurons to trigger action. More specifically, we wait until the background "noise" happens, by chance, to exceed a certain threshold. The RP, on this view, does not indicate a form of unconscious planning: the RP simply records the occasional spikes in neural noise that subjects in Libet's task need to use as cues for action.

Imagine that Bolt is about to demonstrate his speed to an admiring crowd, with no other racers present. We tell him that he won't hear a starter's gun; instead, he should just leave the blocks whenever he feels like it. Given these instructions, we shouldn't be surprised if he ends up going when the background burble of the crowd happens to get a little louder than usual. That does not mean that this background burble is a form of planning, even if it helps to send him out of the blocks. Moreover, the temporal delay between this increase in background noise and Bolt's urge to run does not mean that his urge to run isn't a cause of his running. Once again, he runs because he has the urge, and he has the urge because the crowd gets a little louder. Equally, the fact that Libet found that the RP preceded his subjects' conscious urges to flex their wrists does not mean that those conscious urges were not also efficacious in bringing about movement.

While Libet's RP predicts action a fraction of a second before it occurs, work in the same style from 2008 (work I mentioned at the beginning of the chapter) offers predictive power over a far longer temporal range. John-Dylan Haynes and colleagues asked people to choose to press one of two buttons (one on the left, one on the right, neither of which did anything important) while their brains were monitored using functional magnetic resonance imaging (fMRI): "We found that two brain regions encoded with high accuracy whether the subject was about to choose the left or right response prior to the conscious decision.... The predictive neural information ... preceded the conscious motor decision by up to 10 seconds."[29]

What does this mean?

The "high accuracy" of encoding shouldn't be exaggerated: the experimenters were able to predict whether subjects would choose the right or left button only 60 percent of the time, which means they got it wrong 40 percent of the time. What is more, there are all sorts of reasons why we might find that brain data give us some slight predictive power when it comes to decision-making. Al Mele suggests that perhaps people have slight subconscious biases, which mean they prefer right over left.[30] If those biases are what show up in Haynes's brain scans, then we will be able to use them to predict which button people will choose.

It is bound to be the case, if we are indeed elements of the causal nexus, that our earlier brain states will be of help in predicting what we do. Action takes time and is informed by earlier cognitive processes. As neuroscience advances, we will inevitably find that brain scans illustrate regions that "encode" our future actions. But data like these will not suffice to demonstrate the inefficacy of our conscious decisions, any more than our ability to predict Bolt's lightning departure on the basis of the starter's gun tells us that Usain himself has no control over what he does. Neuroscience has not yet shown freedom to be an illusion.

Further Reading

For classic readings on free will, see:

Gary Watson, ed., *Free Will*, 2nd ed. (Oxford: Oxford University Press, 2003).

A vigorous and entertaining defense of compatibilism can be found in:

Daniel Dennett, *Elbow Room: The Varieties of Free Will Worth Wanting* (Cambridge, MA: MIT Press, 1984).

This chapter takes its title from Dennett's more recent statement of compatibilism, which also includes detailed discussion of Libet:

Daniel Dennett, *Freedom Evolves* (London: Penguin, 2003).

Two important philosophical works on free will, both of which pay close attention to scientific research, are:

Alfred Mele, *Effective Intentions: The Power of Conscious Will* (Oxford: Oxford University Press, 2009).

Robert Kane, *The Significance of Free Will* (Oxford: Oxford University Press, 1996).

Epilogue

The Reach of Science

Scientific Imperialism

The sciences have taught us about the nature of space and the social behavior of microbes. They have illuminated the molecular structure of water and the neural basis of human decisions. How far might the empire of science finally spread? Might the sciences ultimately tell us all there is to know?

In part, these questions turn on the tricky issues of demarcation that we addressed in the first two chapters of this book. Painstaking work in historical archives can help us better understand the causes of wars, the lives of different people at different times, the workings of democratic institutions, and the exercise of political power. Disciplines like history, which are not typically thought of as sciences, regularly uncover important pieces of knowledge.

One might respond by saying that these items of knowledge are inevitably restricted to local times and places. The humanities might illuminate the causes of World War I, or the influence of Martin Luther over the Reformation in German-speaking lands, but only systematic scientific investigation—perhaps of an evolutionary or psychological variety—will tell us about war in general, or religion in general.

That response does not work to elevate the sciences above the disciplines that offer local insight because localization comes by degrees, and many of the sciences offer insight of a localized variety, too. Instead of telling us about the universe in its entirety, evolutionary investigation can reveal the particular pattern of genealogical relationships between primate species, or it might tell us about the functioning of the sperm whale's enormous nose. These claims are confined to particular species, or groups of species, in restricted times and places.

Local knowledge can also be of great value. It would be a mistake to think that because a discipline tells us about general patterns, it must be more useful than a discipline that deals with specificity. This becomes particularly clear when we consider practical responses to problematic situations. Perhaps one of the best-known case studies demonstrating the value of local knowledge comes from the sociologist Brian Wynne's work on the sheep farmers of Northern England and their predicament after the explosion at the Chernobyl nuclear reactor in Ukraine in April 1986.[1]

Local Knowledge

In June 1986, following the detection of radioactive caesium in upland areas of Britain, a ban was placed on the movement and slaughter of sheep in parts of Cumbria. In a small area of Cumbria, that ban lasted far longer than the three weeks initially envisaged by government scientists. Remarkably, restrictions were not lifted entirely until 2012—twenty-six years after the reactor explosion.[2]

Why were scientists' estimates of the time it would take for caesium levels to fall in Cumbria so spectacularly inaccurate? Brian Wynne lays the blame, in part, on the ways in which these scientists overlooked the farmers' valuable local knowledge.

Government scientists' first mistake was to use an inappropriate set of assumptions for understanding how caesium would behave in the upland environment. They thought it would quickly be locked away in the soil, unable to reenter the sheep themselves. Unfortunately, it turned out that these assumptions were true of alkaline clay soils, but not for the acid peaty soils of the Cumbrian uplands. Here, caesium could be constantly recycled, passing from vegetation into lambs, from lambs' manure back into the soil, from the soil back into vegetation, and finally from vegetation back into another generation of lambs.

Scientists then had the idea of spreading a type of clay called bentonite onto the soil, in the hope that it would absorb and capture caesium. They conducted experiments to determine whether this would work and, if so, how much bentonite should be used. Sheep were placed in pens, and the soil in different pens was treated with different amounts of bentonite (and, of course, some contained no bentonite at all). The farmers realized that these experiments would not work, because their sheep did not usually spend their time confined in pens. Instead, they were free to roam around the unfenced fells. If they were confined to pens the sheep would quickly lose condition, and the results of the experiments would be undermined by the sheep's declining health.

Wynne adds that the farmers were also dismayed by scientists' advice to graze their sheep for longer in the valleys, where caesium concentrations were much lower. This, too, ignored knowledge available to the farmers: the supply of grass in the valleys was severely limited. One of the farmers Wynne interviewed told him that if sheep were to be kept there for an extended period of time, the valleys would "be reduced to a desert in days." Government scientists turned out not to have a monopoly on useful knowledge in Cumbria.

The Completeness of Science

There are many facts that are valuable, and which the natural sciences, as conventionally understood, do not deal in. There is no mystery in this: a detailed case study such as Wynne's makes it obvious that facts about how farming is organized in Cumbria—facts that farmers themselves were well placed to impart—are relevant to questions about how radioactive fallout in the area should be managed. Even so, there is nothing to stop scientists of a different sort—social scientists, perhaps—from coming to a detailed understanding of the operation of sheep farming practices in Cumbria. Do we have any reason to think that there are important forms of comprehension that no science could ever achieve?

One of the best-known philosophical thought experiments of the last fifty years, which we owe to the Australian philosopher Frank Jackson, seems to have the consequence that there are some truths that must forever be inaccessible to scientific inquiry.[3] Jackson asks us to imagine a woman called Mary. She is a brilliant scientist. All her life she has studied color and color perception. She knows all about the surface properties of objects, all about the ways in which they reflect light, all about the anatomy of the eye, and all about the ways in which visual information is processed by the brain. In short, she knows all the scientific facts about color and about the perception of color. But Mary has spent her whole life, and has learned all of this science, inside a black and white house, wearing black and white clothes (and gloves), and with no access to windows, or to a mirror. One day she opens the door of her black and white house and steps outside for the first time. She is confronted by a British pillar-box. "Oh!," she says. "I never knew what it was like to see red, but now I do."

Jackson's idea is (or rather, was, because he has changed his mind about the significance of this story) that Mary learns something new when she leaves her house. She learns what it is like to see red. When she was in the house she knew all of the physical facts about redness, and about the perception of red. So if she learns something new, she must learn something nonphysical. She must learn a fact that is not contained within physics.

It is worth pointing out that Jackson's argument has force not merely against the audacious claim that every fact is a fact treated by *physics* but also against the milder claim that every fact is a fact treated by some science or another. A handful of philosophers have been skeptical of the idea that the facts revealed by chemistry, biology, and psychology are all, in some fundamental sense, facts of physics.[4] Those philosophers will not want to say that what Mary learns in her black and white house is all, at root, just physics. After all, our story says that Mary spends her time reading about the physiology of the retina, about the evolution of color vision, and so forth. These are not the sorts of things one learns about in physics lessons. Even if Mary, when confined to her achromatic house, has access to all the journal articles and textbooks she can get her hands on in neuroscience, evolutionary theory, ecology, developmental psychology, anthropology, and so forth, it still seems that she will not know what it is like to see red until she has an encounter with a red object. What she learns on leaving her house does not seem to be a scientific fact of any kind. It seems to follow that there are domains of knowledge that science cannot reach.

One of the most persuasive reactions to Jackson's argument is known as the "ability response." When Mary leaves her black and white house, she learns something she did not know before, and something that was not contained within any of her scientific knowledge. She learns what it is like to see red. But we can conclude that there are nonphysical facts (or nonscientific facts) only if we think that what Mary learns is itself a kind of fact.

Proponents of the ability response—philosophers such as David Lewis, Laurence Nemirow, and Hugh Mellor—have suggested that what Mary learns when she leaves her house is instead a new skill, or ability.[5] She has an encounter for the first time with a red object, and once she has done so she has the ability to recognize more red things, to imagine red things, to recall red things she has seen in the past. Direct experience of a pillar-box does not reveal a special class of fact that the sciences cannot grasp. Instead, it gives Mary a new skill.

One of the nicest features of the ability response is that it gives us a good answer to an awkward question raised by Mary's case. If she is supposed to learn a new fact when she leaves her house—even if it isn't a scientific one—then what explains why that fact isn't one she could have learned by reading a book? One might say that there is something inexpressible about nonscientific facts, which means they cannot be communicated in the normal ways. But this response seems merely to restate our problem: Why does the nonscientific nature of a fact mean that it cannot be written down?

The ability response offers a more satisfactory diagnosis of what happens to Mary. Suppose we alter Jackson's thought experiment, and imagine Bradley, who is mad about bicycles. He has read every book about bikes, he knows all about the history of the Tour de France, and he understands perfectly the physics that explains how bikes work. He has even steeped himself in literature offering tactical advice on how to beat your opponents in a bike race. But Bradley has spent all his life cooped up indoors, without access to a bicycle. Finally, he leaves the confines of his home, to find a brand-new racing machine waiting for him in the driveway. He tries to ride it, and he immediately falls off.

Evidently, for all Bradley's exhaustive book learning, there is something important he does not know about bikes. But we should not say that he is lacking in factual knowledge; instead, he lacks skill. He does not know how to ride a bike. Skills are just the sorts of things that are very difficult to acquire without practice: that is why Bradley could not learn how to ride a bike simply by reading a lot of manuals. Similarly, says the proponent of the ability response, Mary's vast theoretical knowledge of color and color perception was not sufficient for her to know what it is like to see red, because knowledge of what it is like to see red is a skill. The acquisition of that skill demands visual access to red objects, just as learning how to ride a bike demands access to a bike. Scientific papers cannot convey what it is like to see red, because in general it is difficult to acquire skills by reading scientific papers.

If we accept the ability response, then Jackson's thought experiment does not demonstrate the existence of a realm of facts that will forever elude the reach of science. But the ability response amply illustrates the existence of valuable knowledge that the sciences will not capture. We can agree that the sciences will not tell us what it is like to see red. We can move well beyond this basic insight to argue more generally that dry presentations of fact may be less effective than engaging works of fiction in helping us to understand what it is like to suffer from depression, or what it is like to have one's way of life torn up by industrialization. We can agree that fiction offers a form of knowledge, or understanding, that often eludes presentations of the sort one finds in works of psychology or sociology. This does not mean that works of fiction alert us to facts that science cannot grasp. Instead, they give us a way of acquiring and refining important skills.

The generation of scientific knowledge is of course reliant on practical skill, too. Scientists must learn how to design experiments, how to operate equipment, how to interpret data. Although science tells us much that is important, there is no chance that it will ever tell us all that we need to know if we are to understand our world, to live well, and to make wise decisions. The successful mobilization of research for valuable ends demands attention to the sort of local knowledge that science often overlooks. Harnessing scientific work also demands skill, especially skill in judging which pieces of research should be communicated to those with the power to act on them and how that communication should take place. Finally, the significance of scientific research for our self-understanding is not something that leaps immediately from the pages of scientific journal articles: instead, careful interpretation is required when we ask what this work means for freedom of the will, for our moral self-image, for human nature. What is the meaning of science? That is not a question that science will answer on its own.

Further Reading

On local and scientific knowledge, see:

Alan Irwin and Brian Wynne, eds., *Misunderstanding Science?: The Public Reconstruction of Science and Technology* (Cambridge: Cambridge University Press, 1996).

On the relationship between the sciences, and more specifically on the question of whether all science is ultimately just physics, see:

John Dupré, *The Disorder of Things* (Cambridge, MA: Harvard University Press, 1993).

On Mary, see:

Peter Ludlow, Yujin Nagasawa, and Daniel Stoljar, eds., *There's Something About Mary* (Cambridge, MA: MIT Press, 2004).

Notes

Introduction: The Wonder of Science

1. This remark is widely credited to Feynman, but there is no decisive evidence that confirms he really made it.

2. The text of Einstein's letter is taken from D. Howard, "Einstein's Philosophy of Science," in E. N. Zalta, ed., *The Stanford Encyclopedia of Philosophy* (Summer 2010 edition), http://plato.stanford.edu/archives /sum2010/entries/einstein-philscience/.

3. This important theme has been stressed in the work of my Cambridge colleague Hasok Chang.

Chapter 1: How Science Works

1. A. Rosenberg, *Economics—Mathematical Politics or Science of Diminishing Returns?* (Chicago: University of Chicago Press, 1992).

2. W. Dembski and M. Ruse, eds., *Debating Design: From Darwin to DNA* (Cambridge: Cambridge University Press, 2004); S. Sarkar, *Doubting Darwin?: Creationist Designs on Evolution* (Oxford: Blackwell, 2007).

3. See, for example, S. Singh and E. Ernst, *Trick or Treatment: Alternative Medicine on Trial* (London: Bantam Press, 2008).

4. This was the case made in favor of homeopathic remedies at a debate I attended at the Nuffield Council on Bioethics in May 2014.

5. K. Popper, *The Logic of Scientific Discovery* (London: Routledge, 1992); K. Popper, *Unended Quest: An Intellectual Autobiography* (London: Routledge, 1992).

6. D. Gillies, "Lakatos, Popper, and Feyerabend: Some Personal Reminiscences," talk at University College London, Department of Science and Technology Studies, February 28, 2011, http://www.ucl.ac.uk/silva/sts/staff/gillies/gillies_2011_lakatos_popper_feyerabend.pdf.

7. Ibid.

8. Medawar and Bondi are quoted in B. Magee, *Popper* (London: Fontana, 1973), p. 9.

9. Gillies, "Lakatos, Popper, and Feyerabend: Some Personal Reminiscences."

10. K. Popper, "Science: Conjectures and Refutations," in *Conjectures and Refutations: The Growth of Scientific Knowledge* (London: Routledge, 1963), p. 44.

11. Ibid., p. 45.

12. Ibid.

13. This example comes from a *Daily Mail* horoscope published in August 2014: http://www.dailymail.co.uk/home/you/article-1025205/This-weeks-horoscopes-Sally-Brompton.html (accessed August 12, 2014).

14. S. Freud, *The Standard Edition of the Complete Psychological Works,* Vol. 4 (London: Hogarth Press, 1900), p. 150. For further discussion of this example from a philosophical perspective, see A. Grünbaum, "The Psychoanalytic Enterprise in Philosophical Perspective," in C. W. Savage, ed., *Scientific Theories: Minnesota Studies in Philosophy of Science,* Vol. 14 (Minneapolis: University of Minnesota Press, 1990), pp. 41–58.

15. Ibid.; italics in original.

16. For a limpid introduction to the problem of induction, see the first chapter of P. Lipton, *Inference to the Best Explanation, 2nd ed.* (London: Routledge, 2004).

17. Popper, "Science: Conjectures and Refutations," p. 56.

18. I have transcribed Feynman's comments from a video of his lecture, which is available on YouTube (http://youtu.be/EYPapE-3FRw).

19. G. Brumfiel, "Particles Break Light-Speed Limit," *Nature,* September 22, 2011, http://www.nature.com/news/2011/110922/full/news.2011.554.html#update1.

20. Rees's and Weinberg's comments are cited in J. Matson, "Faster-Than-Light Neutrinos? Physics Luminaries Voice Doubts," *Scientific American,* September 26, 2011, http://www.scientificamerican.com/article /ftl-neutrinos/.

21. F. Dyson, A. Eddington, and C. Davidson, "A Determination of the Deflection of Light by the Sun's Gravitational Field, from Observations Made at the Total Eclipse of May 29, 1919," *Philosophical Transactions of the Royal Society of London A* 220 (1920): 332.

22. H. Putnam, "The 'Corroboration' of Theories," in R. Boyd, P. Gasper, and D. Trout, eds., *The Philosophy of Science* (Cambridge, MA: MIT Press, 1991).

23. Popper, *The Logic of Scientific Discovery,* p. 94.

24. Ibid., p. 87.

25. E. Reich, "Embattled Neutrino Project Leaders Step Down", *Nature,* April 2, 2012, http://www.nature.com/news/embattled-neutrino-project -leaders-step-down-1.10371.

26. C. Darwin, *On the Origin of Species* (London: John Murray, 1859).

27. H. Kroto, "The Wrecking of British Science," *The Guardian,* May 22, 2007.

28. Ibid.

29. P. Feyerabend, *Against Method: Outline of an Anarchist Theory of Knowledge* (New York: Verso, 1975), p. 40.

Chapter 2: Is *That* Science?

1. "The Sveriges Riksbank Prize in Economic Sciences in Memory of Alfred Nobel," *Nobelprize.org,* The Official Web Site of the Nobel Prize, n.d., http://www.nobelprize.org/nobel_prizes/economic-sciences/.

2. For an introduction to this research, see D. Kahneman, *Thinking, Fast and Slow* (London: Penguin, 2012).

3. J. Henrich et al., "'Economic Man' in Cross-Cultural Perspective: Behavioral Experiments in 15 Small-Scale Societies," *Behavioral and Brain Sciences* 28 (2005): 795–855.

4. A. Sen, *Poverty and Famines: An Essay on Entitlements and Deprivation* (Oxford: Oxford University Press, 1983).

5. N. Cartwright, *Nature's Capacities and Their Measurement* (Oxford: Oxford University Press, 1989).

6. E. Sober, *The Nature of Selection* (Chicago: University of Chicago Press, 1984), ch. 1.

7. A. Alexandrova, "Making Models Count," *Philosophy of Science* 75 (2008): 383–404.

8. N. Cartwright, "The Vanity of Rigour in Economics: Theoretical Models and Galilean Experiments," in *Hunting Causes and Using Them: Approaches in Philosophy and Economics* (Cambridge: Cambridge University Press: 2007), pp. 217–235.

9. I owe this idea to Hasok Chang.

10. See, for example, M. Ridley, *Evolution*, 3rd ed. (Oxford: Blackwell, 2003); N. Barton et al., *Evolution* (Cold Spring Harbor, NY: Cold Spring Harbor Laboratory Press, 2007).

11. J. Endler, *Natural Selection in the Wild* (Princeton, NJ: Princeton University Press, 1986).

12. The works that are representative of the intelligent-design tradition include W. Dembski and J. Kushiner, eds., *Signs of Intelligence* (Grand Rapids, MI: Brazos Press, 2001), and M. Behe, *Darwin's Black Box*, 2nd ed. (Riverside, NJ: Simon and Schuster, 2006).

13. For a critical survey of a variety of different approaches from creationists and intelligent-design theorists, see R. Pennock, *Tower of Babel* (Cambridge, MA: MIT Press, 1999).

14. The discussion that follows is adapted from a more detailed treatment in my book *Darwin* (London: Routledge, 2007).

15. Behe, *Darwin's Black Box*.

16. K. Miller, "The Flagellum Unspun: The Collapse of 'Irreducible Complexity,'" in W. Dembski and M. Ruse, eds., *Debating Design: From Darwin to DNA* (Cambridge: Cambridge University Press, 2004).

17. E. Sober, "The Design Argument," in W. Mann, ed., *The Blackwell Companion to the Philosophy of Religion* (Oxford: Blackwell, 2004).

18. See also S. Sarkar, *Doubting Darwin? Creationist Designs on Evolution* (Oxford: Blackwell, 2007).

19. "What Is Homeopathy?," British Homeopathic Association, n.d., http://www.britishhomeopathic.org/what-is-homeopathy/.

20. See, for example, C. Weijer, "Placebo Trials and Tribulations," *Canadian Medical Association Journal* 166 (2002): 603–604.

21. R. Smith, "Medical Journals and Pharmaceutical Companies: Uneasy Bedfellows," *BMJ* 326 (2003): 1202.

22. L. Kimber et al., "Massage or Music for Pain Relief in Labour," *European Journal of Pain* 12 (2008): 961–969; E. Ernst, "Does Post-Exercise Massage Treatment Reduce Delayed Onset Muscle Soreness? A Systematic Review," *British Journal of Sports Medicine* 32 (1998): 212–214.

23. The approach that links homeopathic remedies with specific disease conditions has been called "clinical homeopathy," in contrast to the more holistic "classic homeopathy." See Bannerji et al., "Homeopathy for Allergic Rhinitis: Protocol for a Systematic Review," *Systematic Reviews* 3 (2014): 59.

24. "What Is Homeopathy?" British Homeopathic Association, n.d., http://www.britishhomeopathic.org/what-is-homeopathy/.

25. National Health and Medical Research Council (NHMRC) Homeopathy Working Committee, *Effectiveness of Homeopathy for Clinical Conditions: Evaluation of the Evidence,* Optum (2013), p. 8.

26. D. Sackett et al., "Evidence-Based Medicine: What It Is and What It Isn't," *BMJ* 312 (1996): 71.

27. "The Evidence for Homeopathy?" British Homeopathic Association, n.d., http://www.britishhomeopathic.org/evidence/the-evidence-for-homeopathy/.

28. F. Benedetti, *Placebo Effects: Understanding the Mechanisms in Health and Disease* (Oxford: Oxford University Press, 2009).

29. W. Brown, *The Placebo Effect in Clinical Practice* (Oxford: Oxford University Press, 2013), pp. 54–55.

30. R. Hahn, "The Nocebo Phenomenon: Concept, Evidence, and Implications for Public Health," *Preventive Medicine* 26 (1997): 607–611.

31. J. Fournier et al., "Antidepressant Drug Effects and Depression Severity: A Patient-Level Meta-Analysis," *Journal of the American Medical Association* 303 (2010): 47–53.

32. T. Kaptchuk et al., "Placebos Without Deception: A Randomized Controlled Trial in Irritable Bowel Syndrome," *PLOS One* (2010), DOI: 10.1371/journal.pone.0015591.

33. J. Howick et al., "Placebo Use in the United Kingdom: Results from a National Survey of Primary Care Practitioners," *PLOS One* (2013), DOI: 10.1371/journal.pone.0058247.

Chapter 3: The "Paradigm" Paradigm

1. J. Isaac, *Working Knowledge: Making the Human Sciences from Parsons to Kuhn* (Cambridge, MA: Harvard University Press, 2012), p. 211.

2. T. Kuhn, *The Structure of Scientific Revolutions,* 3rd ed. (Chicago: University of Chicago Press, 1996), p. 151.

3. Ibid., p. 181.; see also M. Masterman, "The Nature of a Paradigm," in I. Lakatos and A. Musgrave, eds., *Criticism and the Growth of Knowledge* (Cambridge: Cambridge University Press, 1970).

4. Kuhn, *The Structure of Scientific Revolutions,* p. 175.

5. M.J.S. Hodge, "The Structure and Strategy of Darwin's 'Long Argument,'" *British Journal for the History of Science* 10 (1977): 237–246.

6. E. Lander et al., "Initial Sequencing and Analysis of the Human Genome," *Nature* 409 (2001): 860–921.

7. K. Lindblad-Toh et al., "Genome Sequence, Comparative Analysis and Haplotype Structure of the Domestic Dog," *Nature* 438 (2005): 803–819; Goff et al., "A Draft Sequence of the Rice Genome," *Science* 296 (2002): 92–100; M. Shapiro, "Genomic Diversity and the Evolution of the Head Crest in the Rock Pigeon," *Science* 339 (2013): 1063–1067.

8. Kuhn, *The Structure of Scientific Revolutions.*

9. For detailed discussion of space from a historical and philosophical perspective, see L. Sklar, *Space, Time and Spacetime* (Berkeley: University of California Press, 1974).

10. History as presented in this paragraph is enormously simplified for the purposes of exposition. For a thorough account of the genesis of relativity, see R. Staley, *Einstein's Generation: The Origins of the Relativity Revolution* (Chicago: University of Chicago Press, 2009).

11. T. Kuhn, "Commensurability, Comparability, Communicability," in *The Road Since Structure* (Chicago: University of Chicago Press, 2000), pp. 33–57. On Kuhn's changing ideas about incommensurability, see

H. Sankey, "Kuhn's Changing Concept of Incommensurability," *British Journal for the Philosophy of Science* 44 (1993): 759–774. On incommensurability more generally, see H. Sankey, *The Incommensurability Thesis* (Aldershot, UK: Avebury, 1994).

12. Kuhn, "Commensurability, Comparability, Communicability," in *The Road Since Structure*, p. 48.

13. Kuhn, *The Structure of Scientific Revolutions*.

14. The story of Kuhn's "Aristotle moment" is taken from Isaac, *Working Knowledge*, pp. 211–212.

15. Kuhn, *The Structure of Scientific Revolutions*, p. 134.

16. Ibid., p. 119.

17. I. Hacking, *Representing and Intervening* (Cambridge: Cambridge University Press, 1983).

18. For philosophical discussion of these matters, see A. Byrne and D. Hilbert, eds., *Readings on Color, Vol. 1: The Philosophy of Color* (Cambridge, MA: MIT Press, 1997).

19. Kuhn, *The Structure of Scientific Revolutions*, p. 118.

20. E. Thompson, *Colour Vision: A Study in Cognitive Science and Philosophy of Science* (London: Routledge, 1995).

21. For further discussion of Darwin's views on progress, see my book *Darwin* (London: Routledge, 2007).

22. J. Odling-Smee, K. Laland, and M. Feldman, *Niche Construction: The Neglected Process in Evolution* (Princeton, NJ: Princeton University Press, 2003).

23. Kuhn, "The Road Since *Structure*," in *The Road Since Structure*, p. 104.

24. C. Darwin, *On the Origin of Species* (London: John Murray, 1859).

25. See the "Historical Sketch," which Darwin himself added to later editions of the *Origin,* and which acknowledges some of these precursors.

26. J. Secord, *Victorian Sensation* (Chicago: University of Chicago Press, 2001).

27. P. Bowler, *The Eclipse of Darwinism* (Baltimore: Johns Hopkins University Press, 1983).

28. H. Fleeming Jenkin, "Review of The Origin of Species," *North British Review* 46 (1867): 277–318.

29. T. Lewens, "Natural Selection Then and Now," *Biological Reviews* 85 (2010): 829–835.

30. R. A. Fisher, *The Genetical Theory of Natural Selection* (Oxford: Clarendon Press, 1930).

Chapter 4: But Is It True?

1. For an exploration of a similar position from a Jewish perspective, see P. Lipton, "Science and Religion: The Immersion Solution," in J. Cornwell and M. McGhee, eds., *Philosophers and God: At the Frontiers of Faith and Reason* (London: Continuum, 2009), pp. 1–20.

2. Important papers on underdetermination include L. Laudan, "Demystifying Underdetermination," in C. Wade Savage, ed., *Scientific Theories,* Minnesota Studies in the Philosophy of Science, Vol. 14 (Minneapolis: University of Minnesota Press, 1990), pp. 267–297, and L. Laudan and J. Leplin, "Empirical Equivalence and Underdetermination," *Journal of Philosophy* 88 (1991): 449–472.

3. C. Clark, *The Sleepwalkers: How Europe Went to War in 1914* (London: Penguin, 2013), pp. 47–48.

4. A full transcript of Boas's unpublished lecture is attached as an online appendix to H. Lewis, "Boas, Darwin, Science, and Anthropology," *Current Anthropology* 42 (2001). I am grateful to Jim Moore for first drawing my attention to it.

5. Originally published as P. Duhem, *La Théorie Physique: Son Objet, Sa Structure* (Paris: Chevalier and Rivière, 1906).

6. P. Duhem, *The Aim and Structure of Physical Theory* (Princeton, NJ: Princeton University Press, 1954 [1914]).

7. H.-J. Shin et al., "State-Selective Dissociation of a Single Water Molecule on an Ultrathin MgO Film," *Nature Materials* 9 (2010): 442–447.

8. The preceding exposition is entirely based on H. Chang, *Is Water H_2O?* (Dordrecht, Holland: Springer, 2012).

9. A. Kukla, "Does Every Theory Have Empirically Equivalent Rivals?," *Erkenntnis* 44 (1996): 145.

10. Here I am influenced by K. Stanford, "Refusing the Devil's Bargain: What Kind of Underdetermination Should We Take Seriously?," *Philosophy of Science* 68 [Proceedings] (2001): S1–S12.

11. See, for example, R. Boyd, "On the Current Status of the Issue of Scientific Realism," *Erkenntnis* 19 (1983): 45–90.

12. H. Putnam, *Mathematics, Matter and Method* (Cambridge: Cambridge University Press, 1975), p. 73.

13. Arguments of this form can be found well before Putnam's brief comment of 1975. The Australian philosopher J.J.C. Smart made similar remarks in his 1963 book *Philosophy and Scientific Realism* (London: Routledge, 1963).

14. F. Nietzsche, *The Gay Science*, trans. W. Kaufmann (New York: Random House, 1974 [1887]), Book 3, Section 110.

15. B. van Fraassen, *The Scientific Image* (Oxford: Clarendon Press, 1980), p. 40.

16. P. D. Magnus and C. Callender, "Realist Ennui and the Base Rate Fallacy," *Philosophy of Science* 71 (2004): 320–338.

17. A. Tversky and D. Kahneman, "Evidential Impact of Base Rates," in D. Kahneman, P. Slovic, and A. Tversky, eds., *Judgement Under Uncertainty: Heuristics and Biases* (Cambridge: Cambridge University Press, 1982).

18. On this "minimalist" view of truth, see, among others, P. Horwich, *Truth*, 2nd ed. (Oxford: Oxford University Press, 1998).

19. For a survey and defense of this type of view, see K. Stanford, *Exceeding Our Grasp: Science, History and the Problem of Unconceived Alternatives* (Oxford: Oxford University Press, 2006).

20. L. Boto, "Horizontal Gene Transfer in the Acquisition of Novel Traits by Metazoans," *Proceedings of the Royal Society B* (2014), DOI: 10.1098/rspb.2013.2450.

21. L. Graham et al., "Lateral Transfer of a Lectin-Like Antifreeze Protein Gene in Fishes," *PLoS One* (2008), DOI: 10.1371/journal.pone.0002616.

22. For discussion of these themes, see W. F. Doolittle, "Uprooting the Tree of Life," *Scientific American* 282 (2000): 90–95.

23. Larry Laudan is often credited, probably erroneously, with being an early proponent of the Pessimistic Induction; see L. Laudan, "A Confutation of Convergent Realism," *Philosophy of Science* 48 (1981): 19–49.

24. P. Lipton, "Tracking Track Records," *Aristotelian Society, Supplementary Volume* 74 (2000): 179–205.

25. For significant challenges to this response to the Pessimistic Induction, see K. Stanford, "No Refuge for Realism," *Philosophy of Science* 70 (2003): 913–925, and H. Chang, "Preservative Realism and Its Discontents: Revisiting Caloric," *Philosophy of Science* 70 (2003): 902–912.

26. Stanford, *Exceeding Our Grasp*.

27. My thinking on these matters has been greatly influenced by the unpublished work of my former PhD student Sam Nicholson: S. Nicholson, "Pessimistic Inductions and the Tracking Condition," PhD dissertation, University of Cambridge, 2011.

Chapter 5: Value and Veracity

1. *Shale Gas Extraction in the UK: A Review of Hydraulic Fracturing,* Royal Society/Royal Academy of Engineering (2012): 5, http://www .raeng.org.uk/publications/reports/shale-gas-extraction-in-the-uk.

2. *Scientific Review of the Safety and Efficacy of Methods to Avoid Mitochondrial Disease Through Assisted Conception,* HFEA (2011), http:// www.hfea.gov.uk/docs/2011-04-18_Mitochondria_review_-_final _report.PDF.

3. My account of Lysenko's speech is based on W. deJong Lambert, *The Cold War Politics of Genetic Research: An Introduction to the Lysenko Affair* (Dordrecht, Holland: Springer, 2012).

4. R. M. Young, "Getting Started on Lysenkoism," *Radical Science Journal* 6–7 (1978): 81–105.

5. S. C. Harland, "Nicolai Ivanovitch Vavilov, 1885–1942," *Obituary Notices of the Royal Society* 9 (1954): 259–264.

6. L. Graham, *Science in Russia and the Soviet Union* (Cambridge: Cambridge University Press, 1993), p. 130.

7. For details, see D. Turner, "The Functions of Fossils: Inference and Explanation in Functional Morphology," *Studies in History and Philosophy of Biological and Biomedical Sciences* 31 (2000): 193–212.

8. E. A. Lloyd, *The Case of the Female Orgasm: Bias in the Study of Evolution* (Cambridge, MA: Harvard University Press, 2005).

9. For examples, see Lloyd's website: http://mypage.iu.edu/~ealloyd/.

10. A. Kinsey et al., *Sexual Behavior in the Human Female* (Philadelphia: W. B. Saunders, 1953), p. 164.

11. D. Morris, *The Naked Ape: A Zoologist's Study of the Human Animal* (New York: McGraw-Hill, 1967), p. 79.

12. G. Gallup and S. Suarez, "Optimal Reproductive Strategies for Bipedalism," *Journal of Human Evolution* 12 (1983), p. 195.

13. Lloyd, *The Case of the Female Orgasm*, p. 58.

14. W. H. Masters and V. E. Johnson, *Human Sexual Response* (Boston: Little, Brown 1966), p. 123; Lloyd, *The Case of the Female Orgasm*, p. 182.

15. Lloyd, *The Case of the Female Orgasm*, p. 190.

16. For Lloyd's own resolutely skeptical update, see E. Lloyd, "The Evolution of Female Orgasm: New Evidence and Feminist Critiques," in F. de Sousa and G. Munevar, eds., *Sex, Reproduction and Darwinism* (London: Pickering and Chatto, 2012).

17. D. Puts, K. Dawood, and L. Welling, "Why Women Have Orgasms: An Evolutionary Analysis," *Archives of Sexual Behavior* 41 (2012): 1127–1143.

18. Lloyd, *The Case of the Female Orgasm*, p. 188.

19. R. Levin, "Can the Controversy About the Putative Role of the Human Female Orgasm in Sperm Transport Be Settled with Our Current Physiological Knowledge of Coitus?," *Journal of Sexual Medicine* 8 (2011): 1566–1578.

20. R. Levin, "The Human Female Orgasm: A Critical Evaluation of Its Proposed Reproductive Functions," *Sexual and Relationship Therapy* 26 (2011): 301–314.

21. E. Lloyd, "Pre-Theoretical Assumptions in Evolutionary Explanations of Female Sexuality," *Philosophical Studies* 69 (1993): 139–153.

22. For details on Darwin's early life, see Janet Browne's peerless biography: J. Browne, *Charles Darwin: Voyaging* (London: Pimlico, 2003).

23. The letters from Marx and Engels are both taken from A. Schmidt, *The Concept of Nature in Marx,* trans. B. Fowkes, from the German edition of 1962 (London: New Left Books, 1971).

24. C. Darwin, *On the Origin of Species* (London: John Murray, 1859), p. 108.

25. For details, see J. Odling-Smee, K. Laland, and M. Feldman, *Niche Construction: The Neglected Process in Evolution* (Princeton, NJ: Princeton University Press, 2003).

26. See, for example, R. Levins and R. Lewontin, *The Dialectical Biologist* (Cambridge, MA: Harvard University Press, 1985).

27. The arguments of this section are heavily influenced by the important work of Heather Douglas: H. Douglas, *Science, Policy and the Value-Free Ideal* (Pittsburgh: Pittsburgh University Press, 2009).

28. For a detailed study of responses to allegations of cell phone risk, see A. Burgess, *Cellular Phones, Public Fears and a Culture of Precaution* (Cambridge: Cambridge University Press, 2003).

29. S. John, "From Social Values to *p*-Values: The Social Epistemology of the International Panel on Climate Change," *Journal of Applied Philosophy* (forthcoming).

30. J. O'Reilly, N. Oreskes, and M. Oppenheimer, "The Rapid Disintegration of Projections: The West Antarctic Ice Sheet and the Intergovernmental Panel on Climate Change," *Social Studies of Science* 42 (2012): 709–731.

31. The argument of this section draws on my article "Taking Sensible Precautions," *Lancet* 371 (2008): 1992–1993.

32. C. Sunstein, *Laws of Fear: Beyond the Precautionary Principle* (Cambridge: Cambridge University Press, 2005).

33. *Rio Declaration on Environment and Development,* http://www.unesco.org/education/nfsunesco/pdf/RIO_E.PDF.

34. G. Suntharalingam et al., "Cytokine Storm in a Phase 1 Trial of the Anti-CD28 Monoclonal Antibody TGN1412," *New England Journal of Medicine* 355 (2006): 1018–1028.

35. U. Beck, *Risk Society* (London: Sage, 1992), p. 62; italics in original.

36. S. John, "In Defence of Bad Science and Irrational Policies," *Ethical Theory and Moral Practice* 13 (2010): 3–18.

Chapter 6: Human Kindness

1. C. Darwin, *The Descent of Man* (London: John Murray, 1871), p. 106.

2. M. Ghiselin, "Darwin and Evolutionary Psychology," *Science* 179 (1973): 967.

3. M. Ghiselin, *The Economy of Nature and the Evolution of Sex* (Berkeley: University of California Press, 1974).

4. Darwin, *The Descent of Man,* p. 87.

5. C. Darwin, *On the Origin of Species* (London: John Murray, 1859).

6. R. Alexander, "Evolutionary Selection and the Nature of Humanity," in V. Hösle and C. Illies, eds., *Darwinism and Philosophy* (Notre Dame, IN: University of Notre Dame Press, 2005), p. 309.

7. D. Zitterbart et al., "Coordinated Movements Prevent Jamming in an Emperor Penguin Huddle," *PLoS One* (2011), DOI: 10.1371/journal.pone.0020260.

8. J. Birch, "Gene Mobility and the Concept of Relatedness," *Biology and Philosophy* 29 (2014): 445–476.

9. For further details about social behavior in bacteria, see ibid.; for an elegant discussion of the differences between biological and psychological altruism, see E. Sober and D. Wilson, *Unto Others* (Cambridge, MA: Harvard University Press, 1999).

10. R. Trivers, "The Evolution of Reciprocal Altruism," *Quarterly Review of Biology* 46 (1971): 35–57.

11. S. West, A. Griffin, and A. Gardner, "Social Semantics: Altruism, Cooperation, Mutualism, Strong Reciprocity and Group Selection," *Journal of Evolutionary Biology* 20 (2007): 415–432.

12. R. Dawkins, *The Selfish Gene,* 30th Anniversary Edition (Oxford: Oxford University Press, 2006), p. 4.

13. On the selfish gene theory in evolutionary theory, see A. Gardner and J. Welch, "A Formal Theory of the Selfish Gene," *Journal of Evolutionary Biology* 24 (2011): 1801–1813.

14. R. Dawkins, *The Extended Phenotype* (Oxford: Oxford University Press, 1982); Gardner and Welch, "A Formal Theory of the Selfish Gene."

15. Darwin, *The Descent of Man*.

16. J. Henrich et al., "In Search of Homo Economicus: Behavioral Experiments in 15 Small-Scale Societies," *American Economic Review* 91 (2001): 73–78.

17. R. Frank et al., "Does Studying Economics Inhibit Cooperation?," *Journal of Economic Perspectives* 7 (1993): 159–171; B. Frey and S. Meier, "Are Political Economists Selfish and Indoctrinated? Evidence from a Natural Experiment," *Economic Inquiry* 41 (2003): 448–462.

18. J. Henrich, "Does Culture Matter in Economic Behavior? Ultimatum Game Bargaining Among the Machiguenga of the Peruvian Amazon," *American Economic Review* 90 (2000): 973–979.

19. Darwin, *The Descent of Man*.

20. I owe my understanding of social evolution to several years' worth of tuition from Jonathan Birch.

21. For detailed discussion of green beards, see A. Gardner and S. West, "Greenbeards," *Evolution* 64 (2010): 25–38.

22. L. Keller and K. Ross, "Selfish Genes: A Green Beard in the Red Fire Ant," *Nature* 394 (1998): 573–575.

23. For a summary, see E. Jablonka and M. Lamb, *Evolution in Four Dimensions, rev. ed.* (Cambridge, MA: MIT Press, 2014).

24. See, for example, P. Richerson and R. Boyd, *Not by Genes Alone* (Chicago: University of Chicago Press, 2005).

25. C. el Mouden et al., "Cultural Transmission and the Evolution of Human Behaviour: A General Approach Based on the Price Equation," *Journal of Evolutionary Biology* 27 (2014): 231–241.

Chapter 7: Nature: Beware!

1. S. Pinker, *The Blank Slate: The Modern Denial of Human Nature* (London: Allen Lane, 2002).

2. M. Sandel, *The Case Against Perfection: Ethics in the Age of Genetic Engineering* (Cambridge, MA: Harvard University Press, 2007).

3. L. Kass, "The Wisdom of Repugnance: Why We Should Ban the Cloning of Humans," *Valparaiso University Law Review* 32 (1998): 689.

4. D. Hull, "Human Nature," *PSA: Proceedings of the Biennial Meeting of the Philosophy of Science Association* 2 (1986): 12.

5. M. Ghiselin, *Metaphysics and the Origin of Species* (Albany: SUNY Press, 1997), p. 1.

6. J. Henrich, S. Heine, and A. Norenzayan, "The Weirdest People in the World?," *Behavioral and Brain Sciences* 33 (2010): 61–135.

7. M. Segall, D. T. Campbell, and M. Herskovits, *The Influence of Culture on Visual Perception* (Indianapolis, IN: Bobbs-Merrill, 1966).

8. W.H.R. Rivers, *Reports of the Cambridge Anthropological Expedition to Torres Straits: Physiology and Psychology,* Vol. 2 (Cambridge: Cambridge University Press, 1901).

9. J. Winawer, N. Witthoft, M. Frank, L. Wu, A. Wade, and L. Boroditsky, "Russian Blues Reveal Effects of Language on Color Discrimination," *Proceedings of the National Academy of Sciences* 104 (2007): 7780–7785.

10. For examples and further discussion, see T. Lewens, "Species, Essence and Explanation," *Studies in History and Philosophy of Biological and Biomedical Sciences* 43 (2012): 751–757.

11. S. Okasha, "Darwinian Metaphysics: Species and the Question of Essentialism," *Synthese* 131 (2002): 191–213.

12. Darwin, *On the Origin of Species* (London: John Murray, 1859).

13. E. Machery, "A Plea for Human Nature," *Philosophical Psychology* 21 (2008): 321–329.

14. T. Lewens, "Human Nature: The Very Idea," *Philosophy and Technology* 25 (2012): 459–474.

15. B. Sinervo and C. M. Lively, "The Rock-Paper-Scissors Game and the Evolution of Alternative Male Strategies," *Nature* 380 (1996): 240–243.

16. M. Tomasello, *The Cultural Origins of Human Cognition* (Cambridge, MA: Harvard University Press, 1999).

17. C. Heyes, "Grist and Mills: On the Cultural Origins of Cultural Learning," *Philosophical Transactions of the Royal Society B* 367 (2012): 2181–2191.

18. C. Heyes, "Causes and Consequences of Imitation," *Trends in Cognitive Sciences* 5 (2001): 253–261.

19. J. Hope, "Inability to Recognise People's Faces Is Inherited," *Daily Mail*, May 8, 2014, http://www.dailymail.co.uk/health/article-2622909 /Find-hard-place-face-Its-genes-Inability-recognise-people-inherited -study-says.html.

20. For a lucid introduction to the meaning of heritability, see E. Sober, "Separating Nature and Nurture," in D. Wasserman and R. Wachbroit, eds., *Genetics and Criminal Behavior: Methods, Meanings and Morals* (Cambridge: Cambridge University Press, 2001), pp. 47–78.

21. The remainder of this section draws on a blog post I have written for Cambridge's Centre for Research in Arts, Social Sciences, and Humanities.

22. P. Wintour, "Genetics Outweighs Teaching, Gove Adviser Tells His Boss," *The Guardian*, October 11, 2013, http://www.theguardian.com /politics/2013/oct/11/genetics-teaching-gove-adviser.

23. R. Plomin and K. Asbury, *G Is for Genes: The Impact of Genetics on Education and Achievement* (Oxford: Wiley-Blackwell, 2013).

24. Quoted in T. Helm, "Michael Gove Urged to Reject 'Chilling Views' of His Special Adviser," *The Observer*, October 12, 2013, http://www .theguardian.com/politics/2013/oct/12/michael-gove-special-adviser.

25. P. Wilby, "Psychologist on a Mission to Give Every Child a Learning Chip," *The Guardian*, February 18, 2014, http://www.theguardian .com/education/2014/feb/18/psychologist-robert-plomin-says-genes -crucial-education.

26. S. Atran et al., "Generic Species and Basic Levels: Essence and Appearance in Folk Biology," *Journal of Ethnobiology* 17 (1997): 17–43.

27. S. Gelman and L. Hirschfeld, "How Biological Is Essentialism?," in S. Atran and D. Medin, eds., *Folkbiology* (Cambridge, MA: MIT Press, 1999), pp. 403–445.

28. I owe this thought to Rae Langton.

29. L. Kass, "Ageless Bodies, Happy Souls: Biotechnology and the Pursuit of Perfection," *The New Atlantis* 1 (2003): 9–28.

30. Kass, "The Wisdom of Repugnance," pp. 689–690.

31. Ibid., p. 691.

32. Ibid.

Chapter 8: Freedom Dissolves?

1. C. Soon et al., "Unconscious Determinants of Free Decisions in the Human Brain," *Nature Neuroscience* 11 (2008): 543–545.

2. S. Harris, *Free Will* (New York: The Free Press, 2012).

3. T. Chivers, "Neuroscience, Free Will and Determinism," *Daily Telegraph*, October 12, 2010, http://www.telegraph.co.uk/science/8058541 /Neuroscience-free-will-and-determinism-Im-just-a-machine.html.

4. M. Gazzaniga, "Free Will Is an Illusion, But You're Still Responsible for Your Actions," *Chronicle of Higher Education*, March 18, 2012, http:// chronicle.com/article/Michael-S-Gazzaniga/131167.

5. Soon et al., "Unconscious Determinants of Free Decisions in the Human Brain."

6. B. Libet et al., "Time of Conscious Intention to Act in Relation to Onset of Cerebral Activity (Readiness Potential): The Unconscious Initiation of a Freely Voluntary Act," *Brain* 106 (1983), 623–642; B. Libet, "Do We Have Free Will?," *Journal of Consciousness Studies* 6 (1999): 54.

7. J. Coyne, "You Don't Have Free Will," *Chronicle of Higher Education*, March 18, 2012, http://chronicle.com/article/Jerry-A-Coyne/131165/.

8. For an important defense of the role of indeterminism in free will, see R. Kane, *The Significance of Free Will* (Oxford: Oxford University Press 1996).

9. D. Dennett, *Elbow Room: The Varieties of Free Will Worth Wanting* (Cambridge, MA: MIT Press, 1984).

10. D. Wooldridge, *The Machinery of the Brain* (New York: McGraw -Hill, 1963), pp. 82–83.

11. F. Keijzer, "The Sphex Story: How the Cognitive Sciences Kept Repeating an Old and Questionable Anecdote," *Philosophical Psychology* 26 (2013): 502–519.

12. H. G. Wells, J. S Huxley, and G. P. Wells, *The Science of Life, Vol. 2* (Garden City, NY: Doubleday, Doran and Company, 1938).

13. J. Fabre, *Souvenirs Entomologiques* (Paris: Librarie Ch. Delagrave, 1879); J. Fabre, *The Hunting Wasps* (New York: Dodd, Mead and Company, 1915).

14. Fabre, *The Hunting Wasps*, p. 78.

15. J. Brockmann, "Provisioning Behavior of the Great Golden Digger Wasp, *Sphex ichneumoneus,*" *Journal of the Kansas Entomological Society* 58 (1985): 631–655.

16. Keijzer, "The Sphex Story."

17. See R. Lurz, *Mindreading Animals: The Debate Over What Animals Know About Other Minds* (Cambridge, MA: MIT Press, 2011).

18. See, for example, D. Kahneman, P. Slovic, and A. Tversky, eds., *Judgement Under Uncertainty: Heuristics and Biases* (Cambridge: Cambridge University Press, 1982).

19. For discussion of these points in the context of health and safety legislation, see C. Sunstein, *Laws of Fear: Beyond the Precautionary Principle* (Cambridge: Cambridge University Press, 2005).

20. For examples of this claim, see the references in E. Nahmias et al., "Surveying Freedom: Folk Intuitions About Free Will and Moral Responsibility," *Philosophical Psychology* 18 (2005): 561–584.

21. Nahmias et al., "Surveying Freedom."

22. Jonathan Birch pointed this problem out to me.

23. Coyne, "You Don't Have Free Will."

24. Libet et al., "Time of Conscious Intention to Act."

25. Examples include T. Bayne, "Libet and the Case for Free Will Scepticism," in R. Swinburne, ed., *Free Will and Modern Science* (Oxford: Oxford University Press/British Academy, 2011); D. Dennett, *Freedom Evolves* (London: Penguin, 2003); and A. Mele, *Effective Intentions: The Power of Conscious Will* (Oxford: Oxford University Press, 2009).

26. I owe this important point to the philosopher Al Mele's detailed discussion of works by Libet and others.

27. J. Trevena and J. Miller, "Brain Preparation Before a Voluntary Action: Evidence Against Unconscious Movement Initiation," *Consciousness and Cognition* 19 (2010): 447–456.

28. A. Schurger et al., "An Accumulator Model for Spontaneous Neural Activity Prior to Self-Initiated Movement," *Proceedings of the National Academy of Sciences* 109 (2012): E2904–E2913.

29. Soon et al., "Unconscious Determinants of Free Decisions in the Human Brain."

30. A. Mele, "The Case Against the Case Against Free Will," *Chronicle of Higher Education,* March 18, 2012, http://chronicle.com/article /Alfred-R-Mele-The-Case/131166/.

Epilogue: The Reach of Science

1. See, for example, B. Wynne, "Misunderstood Misunderstandings," *Public Understanding of Science* 1 (1992): 281–304.

2. BBC News, "Post-Chernobyl Disaster Sheep Controls Lifted on Last UK Farms," June 1, 2012, http://www.bbc.co.uk/news/uk-england -cumbria-18299228.

3. F. Jackson, "Epiphenomenal Qualia," *Philosophical Quarterly* 32 (1982): 127–136; F. Jackson, "What Mary Didn't Know," *Journal of Philosophy* 83 (1986): 291–295.

4. See, for example, T. Crane and D. H. Mellor, "There Is No Question of Physicalism," *Mind* 99 (1990): 185–206, and J. Dupré, *The Disorder of Things* (Cambridge, MA: Harvard University Press, 1993).

5. D. Lewis, "What Experience Teaches," in W. Lycan, ed., *Mind and Cognition* (Oxford: Blackwell, 1990); L. Nemirow, "Physicalism and the Cognitive Role of Acquaintance," in W. Lycan, ed., *Mind and Cognition* (Oxford: Blackwell, 1990); D. H. Mellor, "Nothing Like Experience," *Proceedings of the Aristotelian Society* 93 (1992–1993): 1–16.

Index

TIM LEWENS is a professor of philosophy of science at Cambridge University and a fellow of Clare College. As well as having written a number of titles on biology and bio-ethics, he has written for the *London Review of Books* and the *Times Literary Supplement,* and has won prizes for both his teaching and his publications. He lives in Barton, near Cambridge, UK.

Photo by Emma Gilby